Physics for the Millions

By
Joseph M. Brown

Basic Research Press

Physics for the Millions

©
2013

By
Joseph M. Brown
First Edition
First Impression

ISBN: 978-0-9883180-1-4
Published By
Basic Research Press
120 East Main Street
Starkville, MS 39759
United States of America
basicresearchpress.com

Acknowledgement

I thank Brittney Marie Patterson for transcribing and illustrating this book.

Joseph M. Brown

Physics for the Millions

Contents

Abstract..v
1. What Is the Smallest Thing in the World? The Brutino....................1
2. Practically All of the Universe Is Simply a Gas of Brutinos- The Remainder Is Neutrinos and They Are Made of Brutinos..24
3. Neutrinos Are Made of Brutinos..46
4. The Proton and Electron are Made of Neutrinos.............................64
5. The Electrostatic Force Is the Glue Binding an Electron and a Proton to Produce a Hydrogen Atom....................77
6. The Proton Electrostatic Field Mixed With the Electron Electrostatic Field Produces Gravity.................................87
7. Hydrogen Stars...92
8. The Neutron and the More Massive Atoms.....................................95
9. Neutron Stars and the Big-Big-Bang Theory..................................101
10. Photons Move Almost Everything We See.....................................113
11. Summary...131
    Bibliography...133
    Index..134

Abstract

Brutinos are the smallest particles in the world. Brutinos make up a gas that pervades the universe. Condensation of the brutino gas produces neutrinos which also pervade the universe. A proton is made of one neutrino. When a proton is made, an electron is simultaneously produced. The result is a hydrogen atom. Hydrogen atoms are continually made throughout the universe. Hydrogen atoms accumulate because of their gravitational fields. Continuous accumulation makes hydrogen stars. Large hydrogen stars have large gravitational fields and they develop pressures large enough to transmute a hydrogen atom into a neutron. Protons and neutrons are nucleons. The neutrinos making the protons in a pair of nucleons attract each other and make larger atoms. Hydrogen stars keep growing and start producing more and more large atoms. As stars continue to grow, the electronic structure collapses and the star becomes a neutron star. A neutron star continues to grow and gravity becomes so large that the nuclear structure collapses. The giant neutron star explodes and produces the items which exist throughout the universe today.

1. What is the Smallest Thing in the World? The Brutino

Just what is the smallest thing in the world? Is it a grain of sand, a flake of powder, an atom, a proton, an electron? Is there something smaller than an electron? Is there a small piece of mass from which all material things are made? What is a photon made of? Have you ever heard of a neutrino which travels at the speed of light and can whisk through the earth entering at the United States of America and exiting at China -- without ever hitting an atom? What is a neutrino made of? Is there an ultimately small particle, or something else, from which everthing is made? This book addresses these questions.

While discussing small things we find it convenient to use exponents of 10 for measurements and other descriptions.

$1,000,000 = 10^6$	$1/10 = 10^{-1}$
$100,000 = 10^5$	$1/100 = 10^{-2}$
$10,000 = 10^4$	$1/1000 = 10^{-3}$
$1000 = 10^3$	$1/10,000 = 10^{-4}$
$100 = 10^2$	$1/100,000 = 10^{-5}$
$10 = 10^1$	$1/1,000,000 = 10^{-6}$
$1 = 10^0$	$1/10,000,000 = 10^{-7}$

Physics for the Millions

To answer what the smallest thing is we need magnifiers. In fact, we need imaginary magnifiers which will magnify much greater than anything real.

Here we show a magnifying glass.

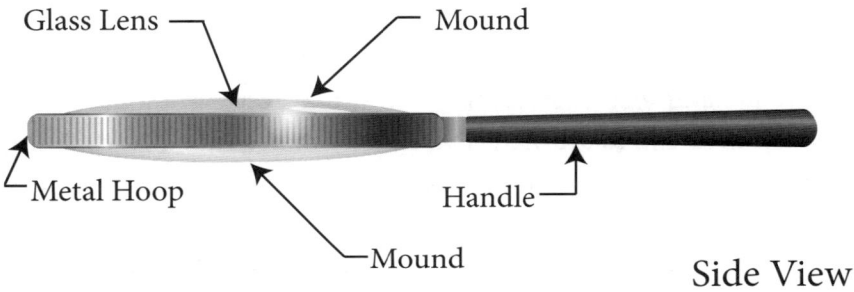

Side View

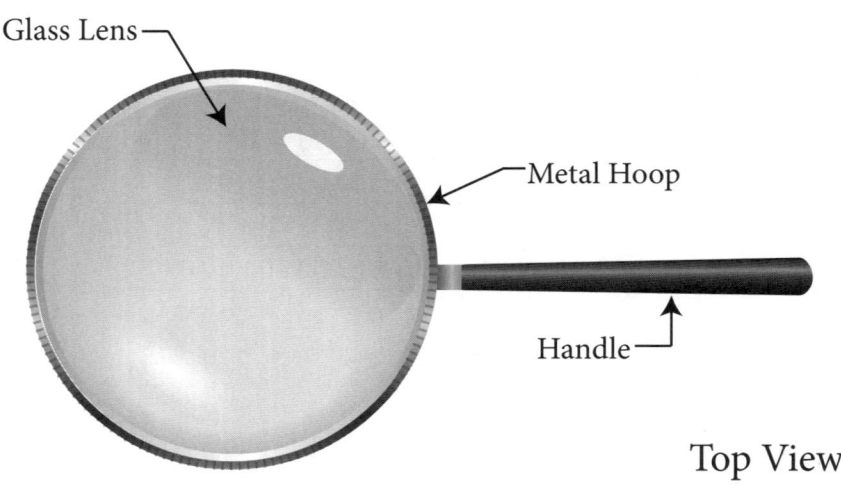

Top View

What is the Smallest Thing in the World? The Brutino

We see a penny magnified by a factor of 2, i.e., two times or 2X.

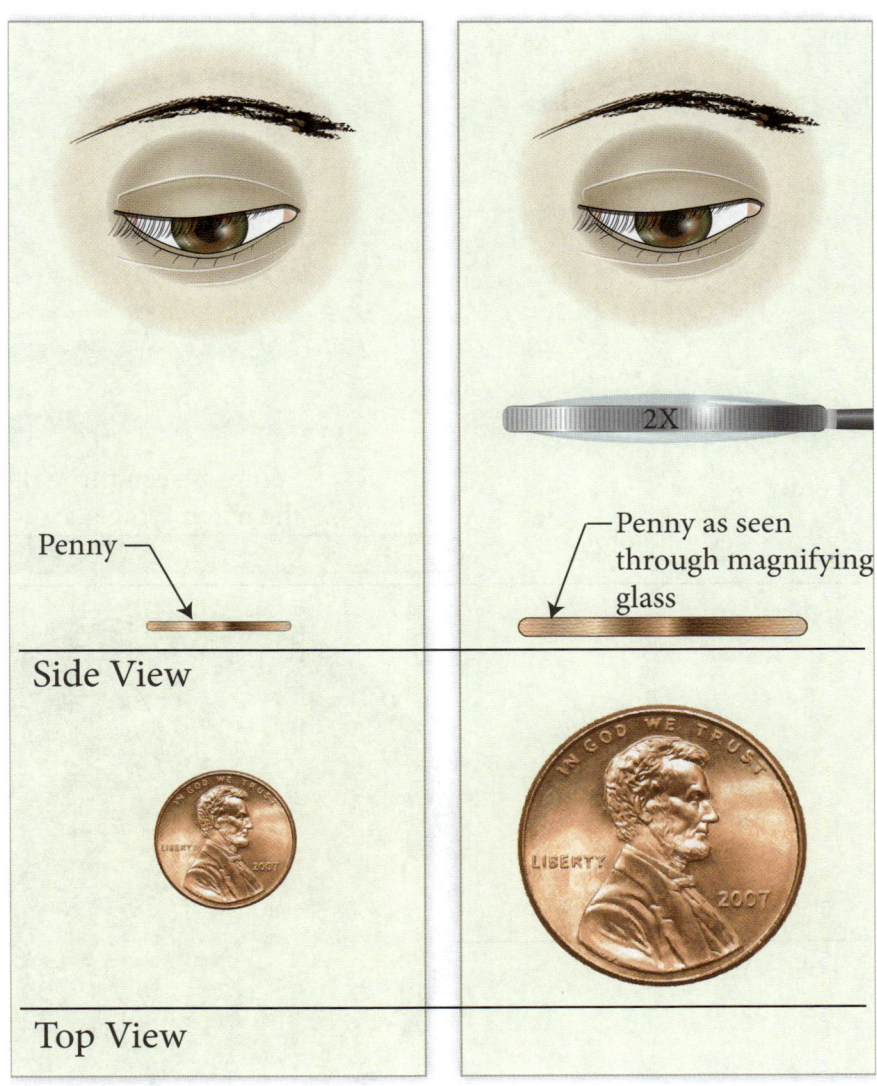

3

Now we magnify a penny by two 2X magnifying glasses in series.

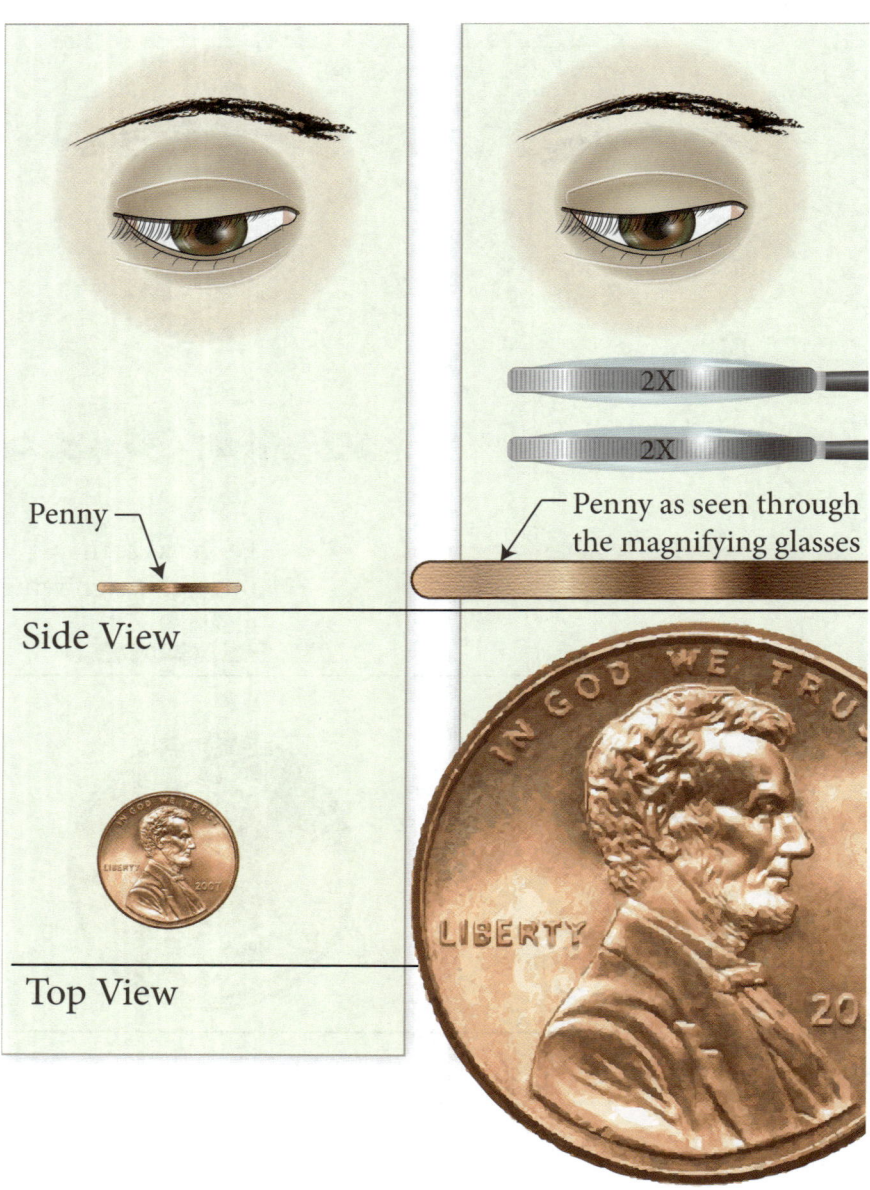

What is the Smallest Thing in the World? The Brutino

Magnifying a penny 10X

Magnifying a penny using two 10X magnifying glasses in series to give a magnification of 10X times 10X or 100X,

Magnification = 10X

Magnification = 10X times 10X = 100X = 10^2X

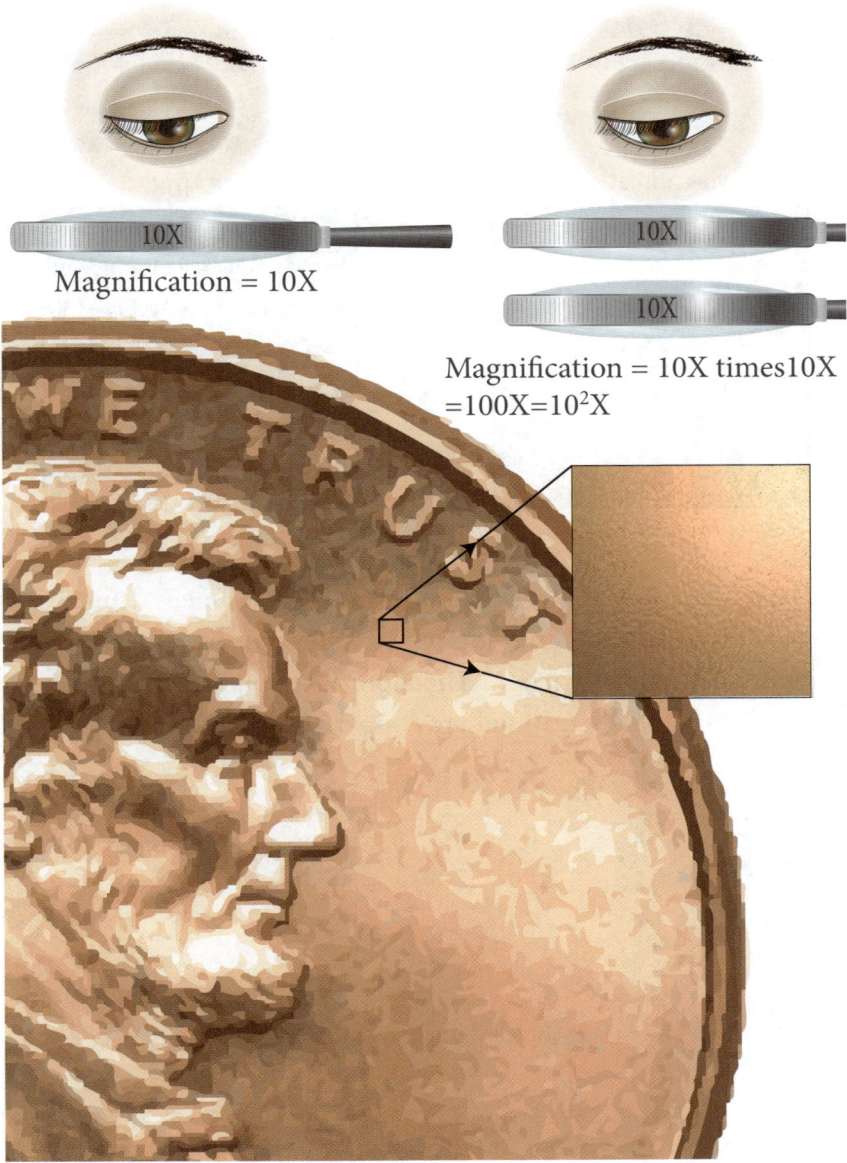

Here we show 100X and 10^8X magnification of a penny.

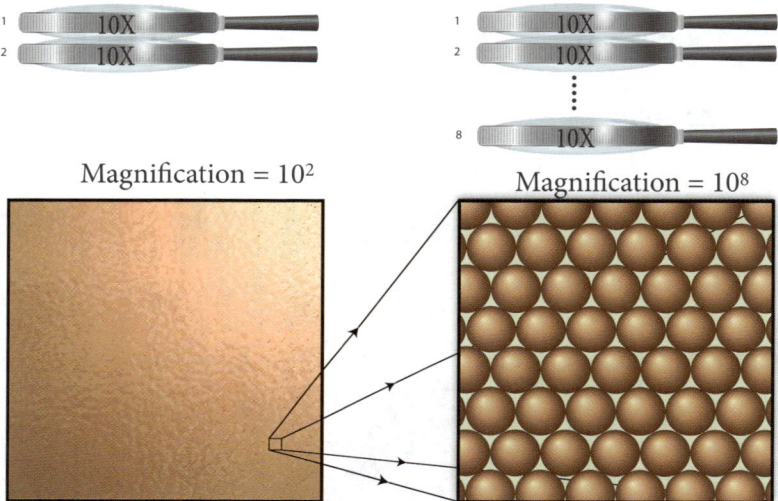

At a magnification of 10^8 we see the atoms of copper. Thus, when magnifying a penny a hundred million times (i.e., by 10^8), we see the indivudiual atoms of copper. The atoms are spherical and they are packed tightly together. What are copper atoms made of?

What is the Smallest Thing in the World? The Brutino

    Instead of delving further into copper, let's switch to water. Water is made of hydrogen and oxygen. We'll want to find out what hydrogen is -- since hydrogen is the simplest of all atoms.

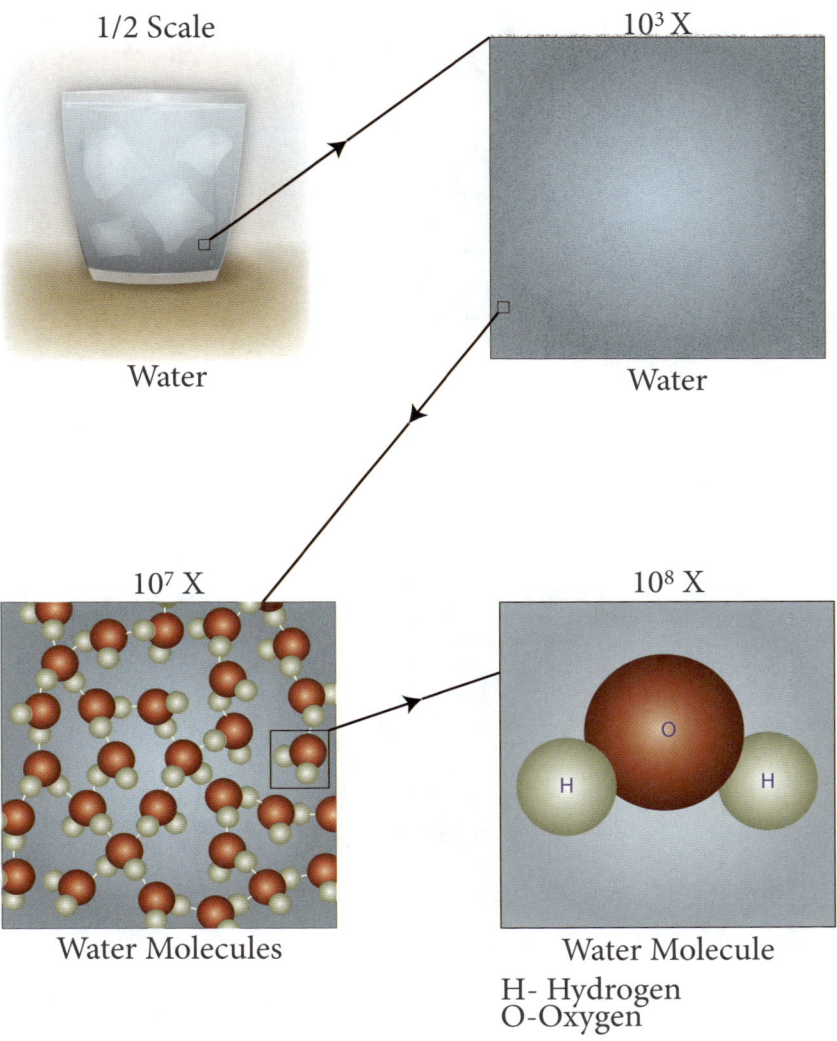

Physics for the Millions

Below is a schematic of the hydrogen atom showing orbit size. The electron and proton are not at the same scale as the orbital radius.

Hydrogen
10^8 X

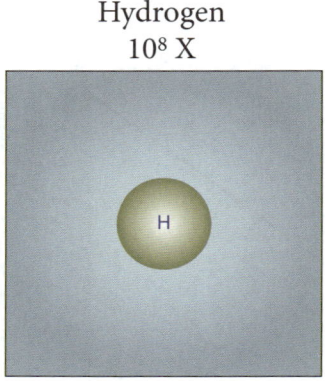

Magnification
10^9X

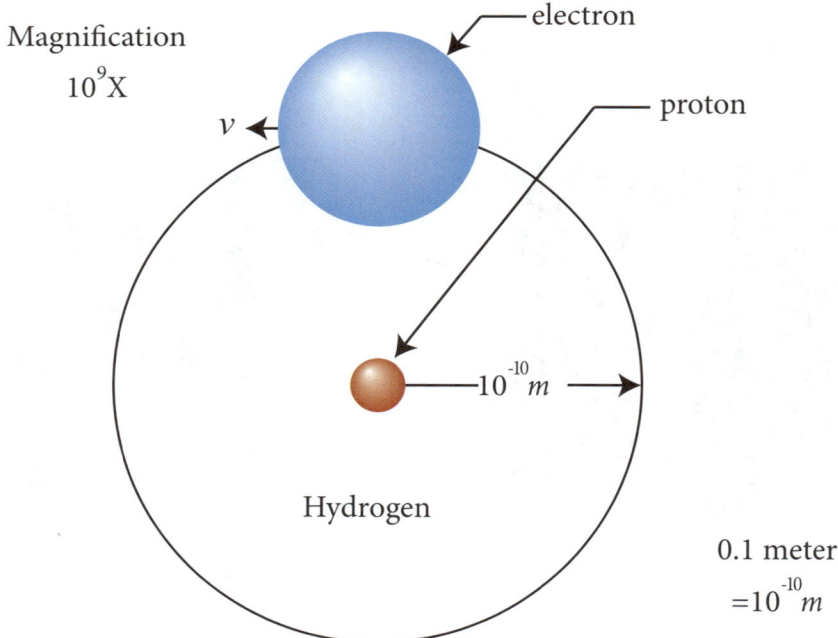

0.1 meter
$=10^{-10} m$

What is the Smallest Thing in the World? The Brutino

Hydrogen atom showing electron and proton orbital radii and velocities.

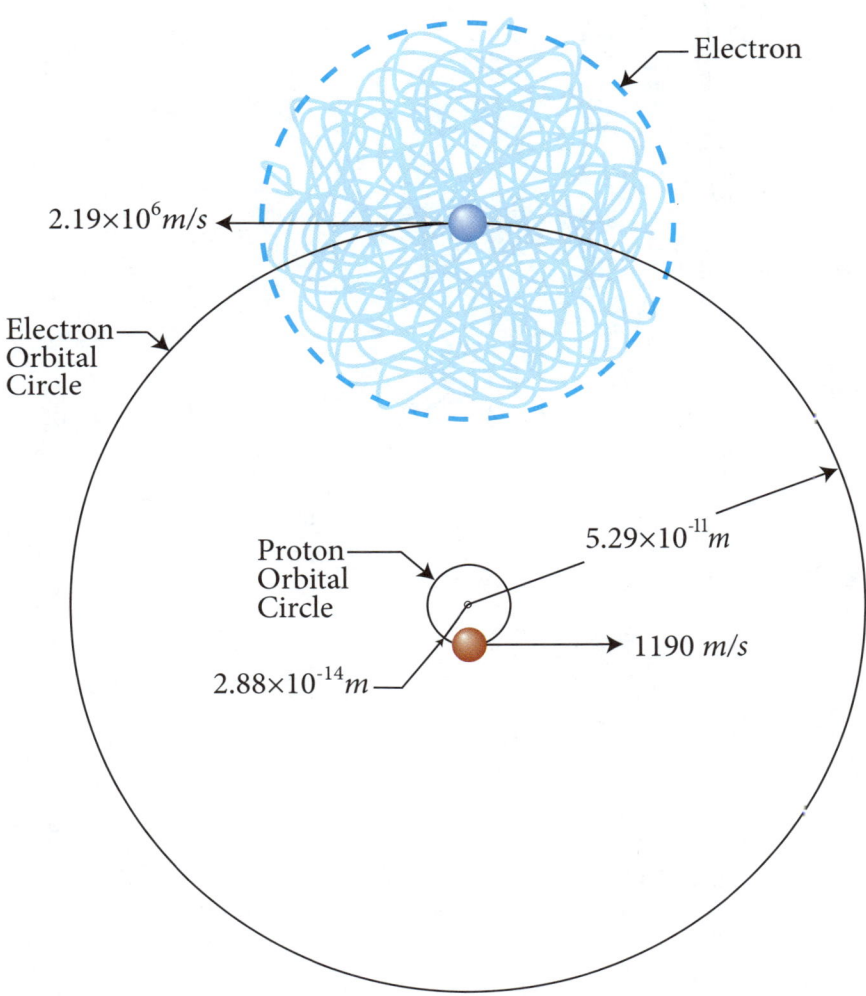

Hydrogen atom magnified, showing the neutrinos which make the electron and proton.

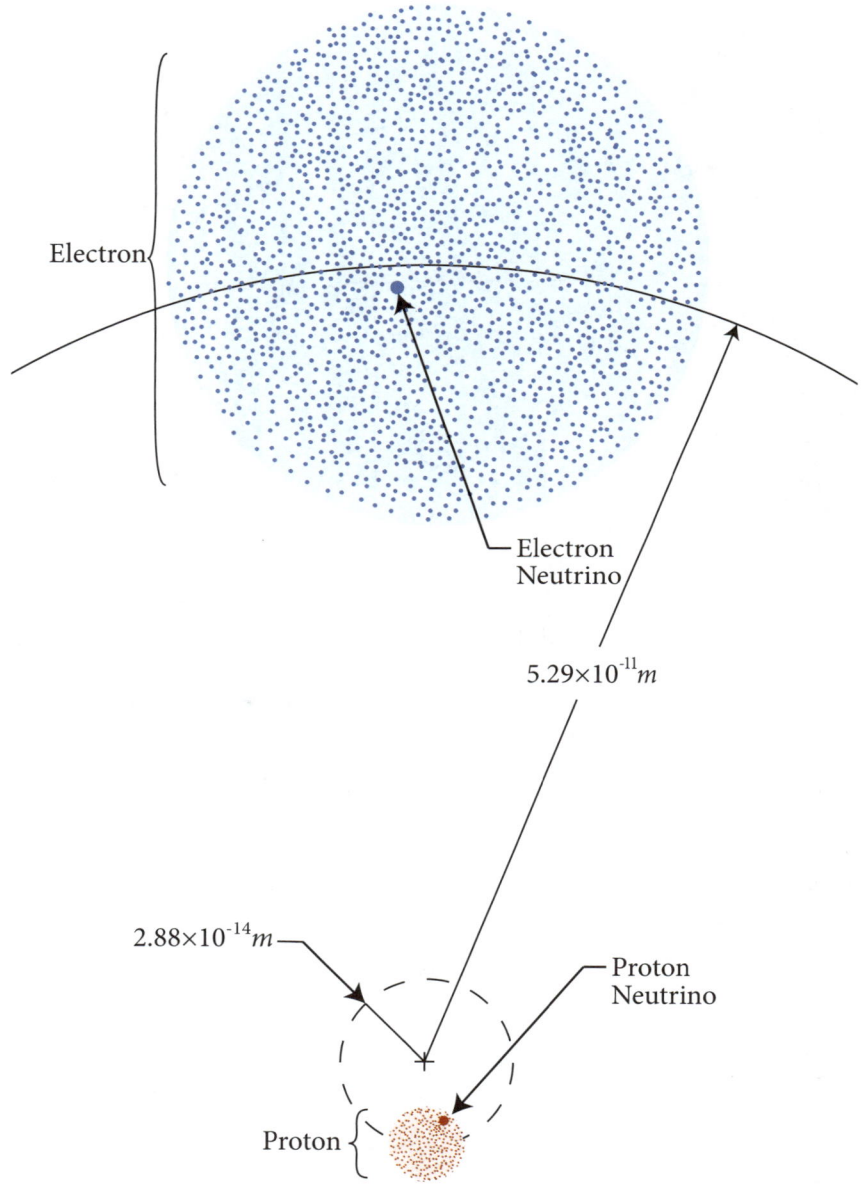

What is the Smallest Thing in the World? The Brutino

Hydrogen atom showing speeds and that the electron and proton are made of orbiting neutrinos moving at the speed of light. The electron is simplified in this illustration.

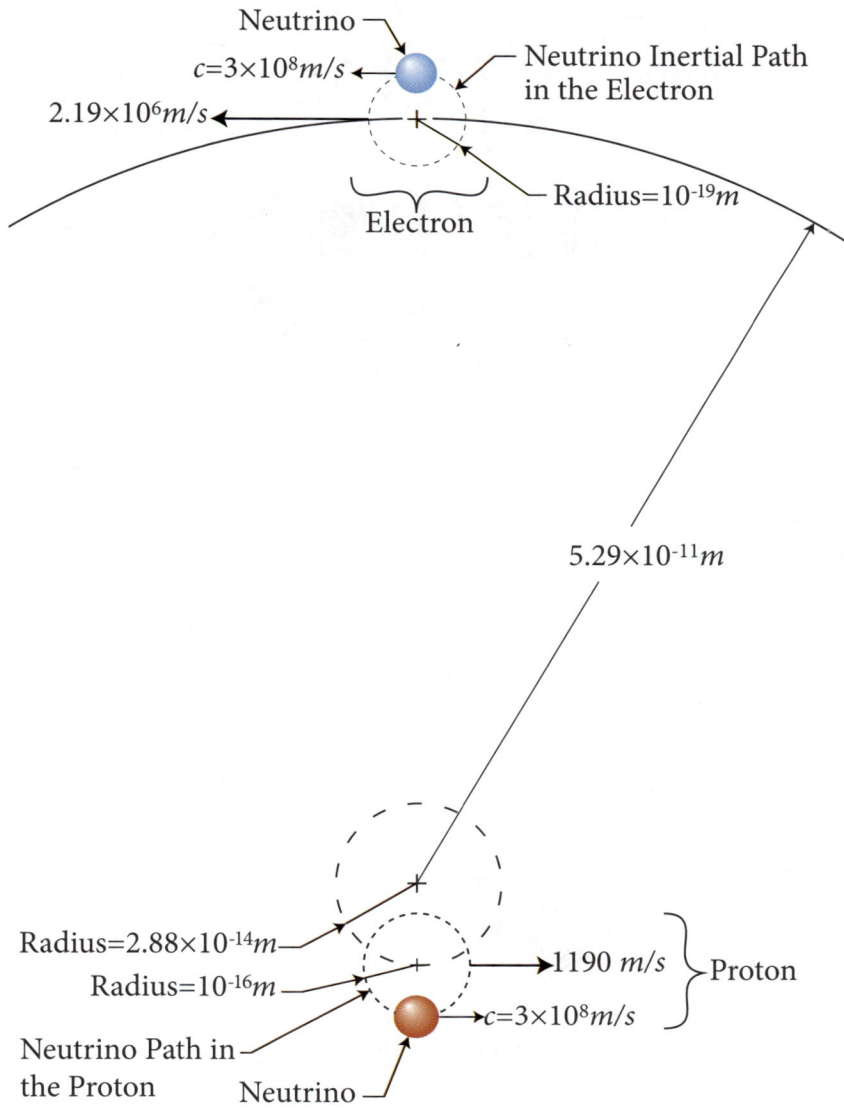

Proton-Sized Neutrino

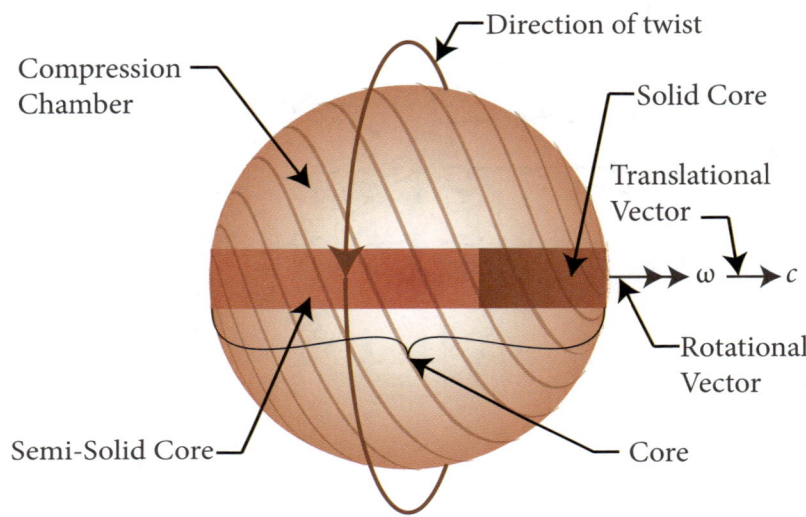

Proton neutrino and its orbital path.

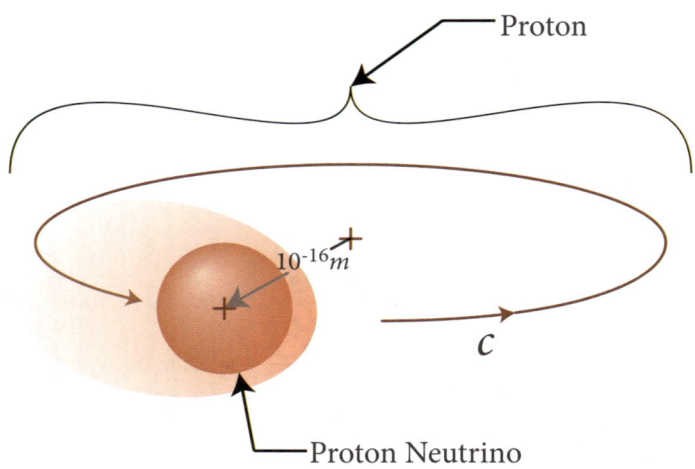

What is the Smallest Thing in the World? The Brutino

Electron-Sized Neutrino

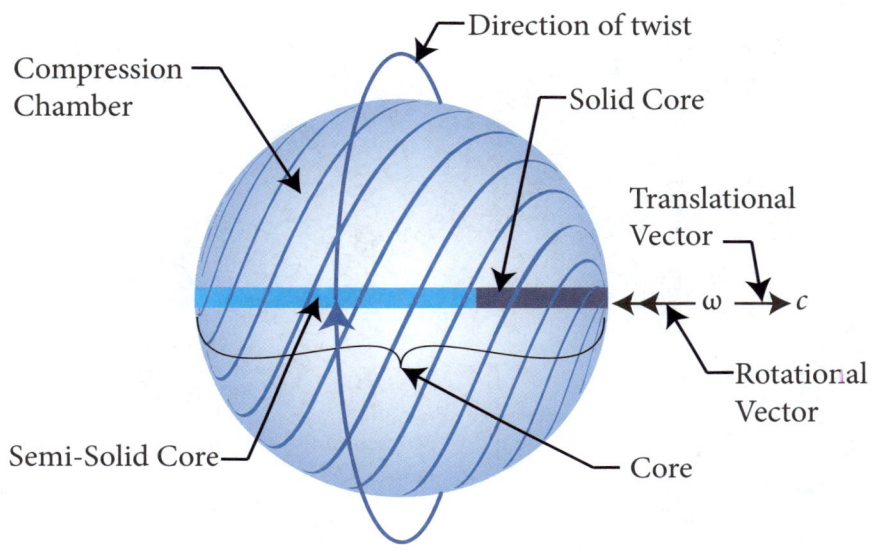

Electron neutrino and its orbital path.

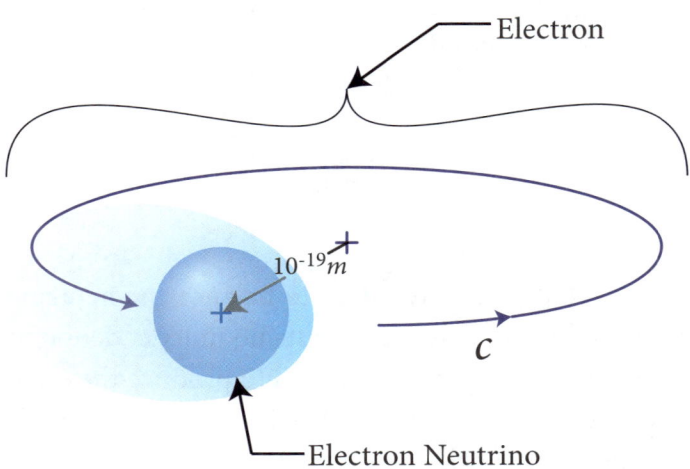

The Paths of the Electron

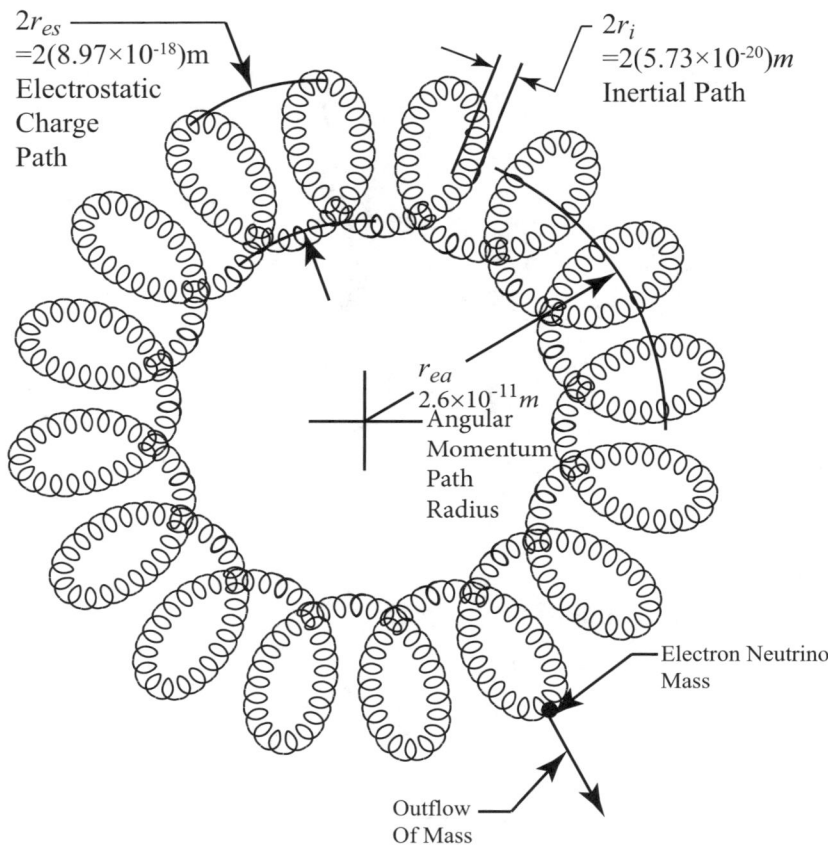

The inertial path is required to balance the neutrino thrust which thrust is the same for every neutrino independent of its mass. The electrostatic charge path is required to produce the electrostatic field. The angular momentum path is required to produce the angular momentum $\hbar/2$.

The Hydrogen Atom

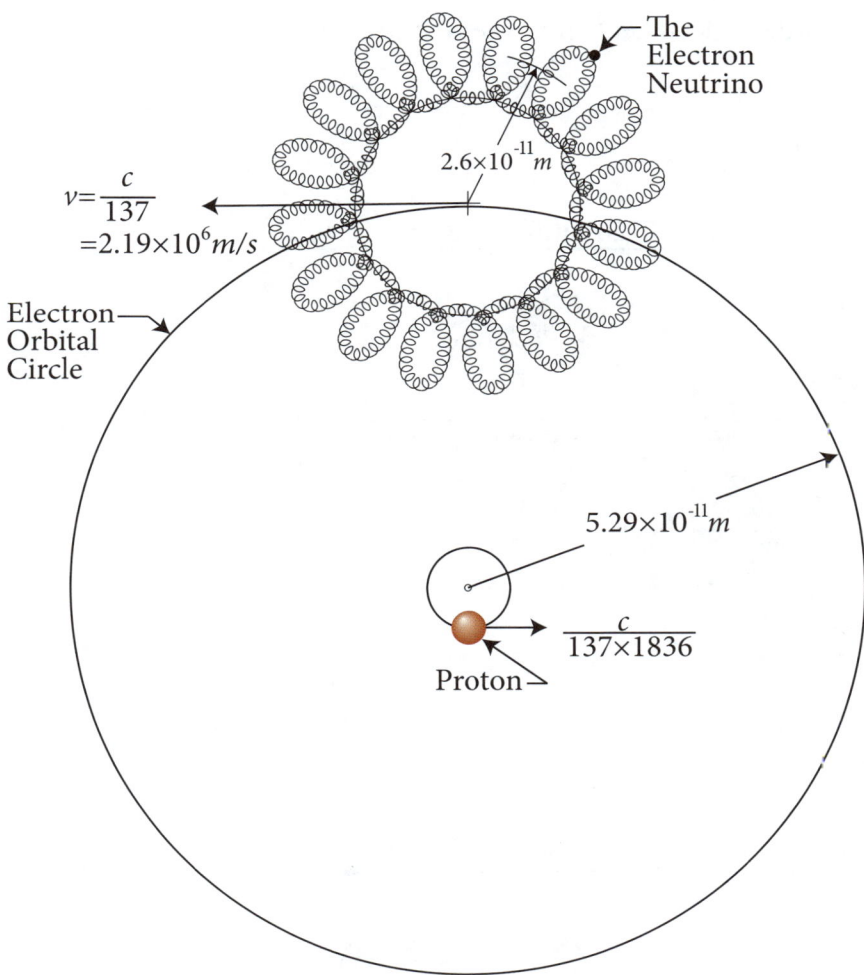

By analogy with the proton, the electron is considered to be the electron neutrino and the inertial path, i.e., the loop with a radius of $5.73 \times 10^{-20} m$.

Notice the rotational vector such as in the figure on Page 12 showing the proton-sized neutrino. The rotation is called twist.

Right- and Left- Hand Twist

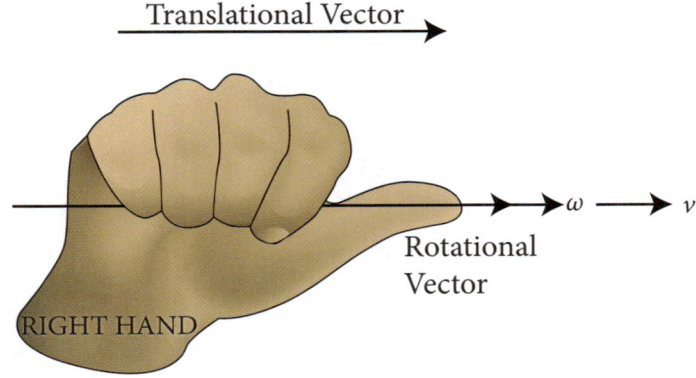

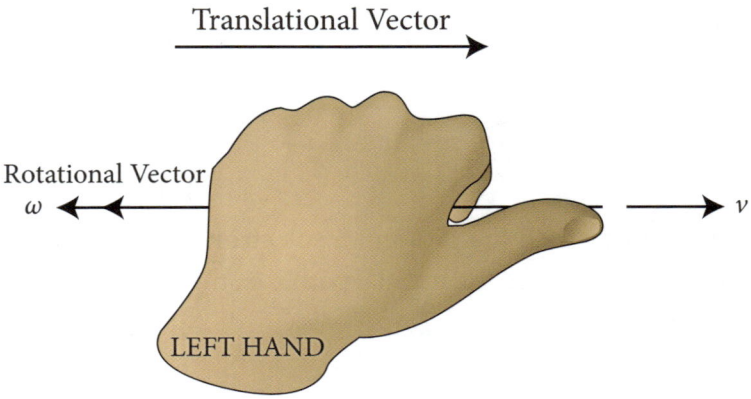

What is the Smallest Thing in the World? The Brutino

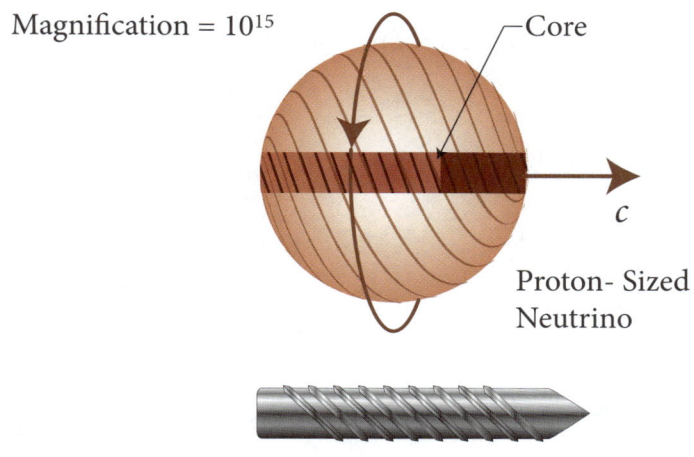

Twisting a right-hand screw in the direction indicated by the fingers as well as the double headed vector will advance the screw in the direction of the velocity vector *c*.

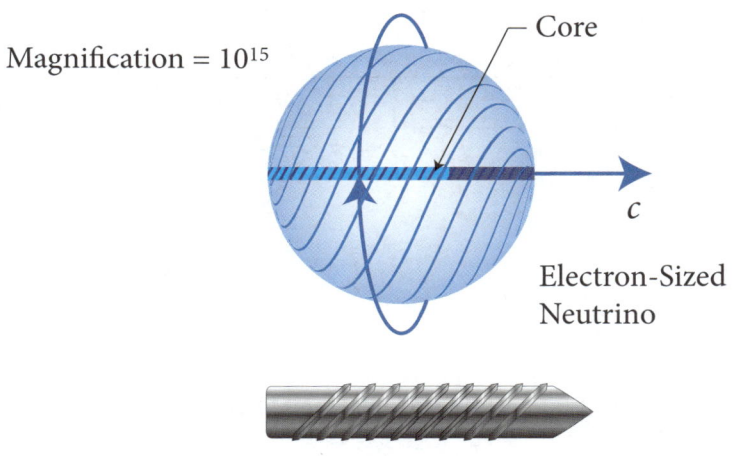

Twisting a left-handed screw in the direction indicated by the fingers as well as the double-headed vector will advance the screw in the direction of the velocity vector *c*.

Details of the Neutrino

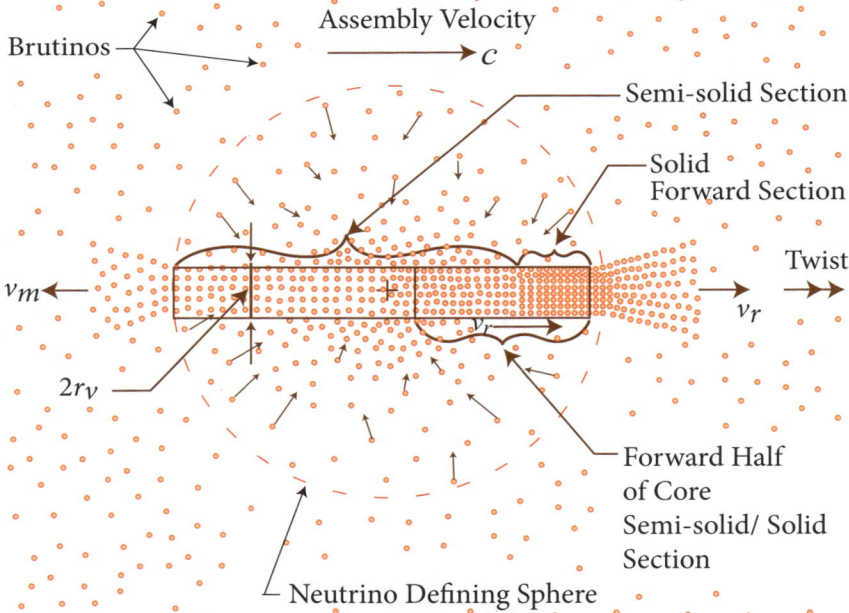

What is the Smallest Thing in the World? The Brutino

More details of the neutrino

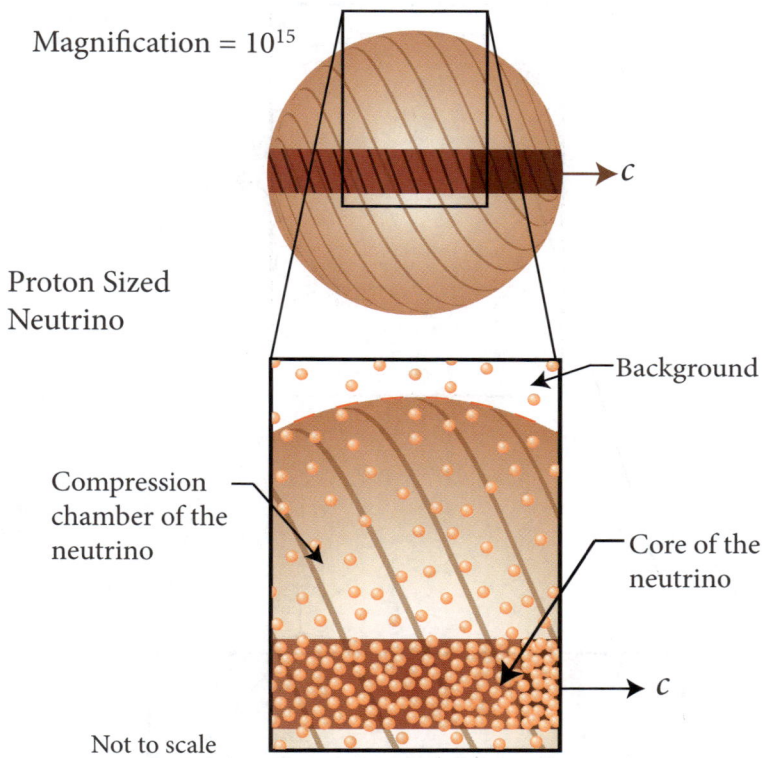

Continual magnification of the background and the neutrino

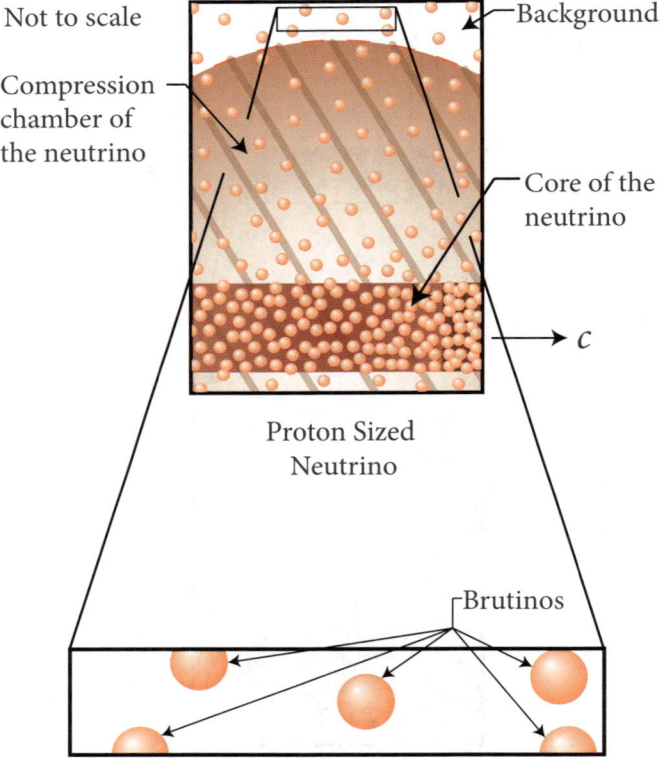

What is the Smallest Thing in the World? The Brutino

Further magnification of the background

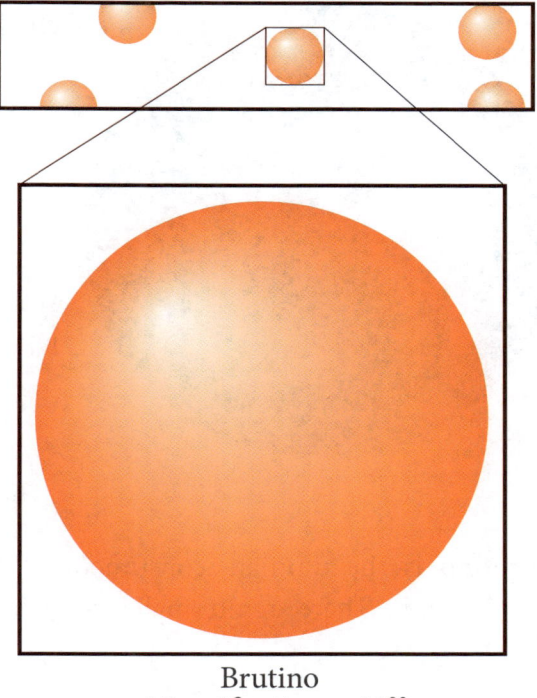

Brutino
Magnification = 10^{33}

The brutino
   The gas particle making up everything in the universe

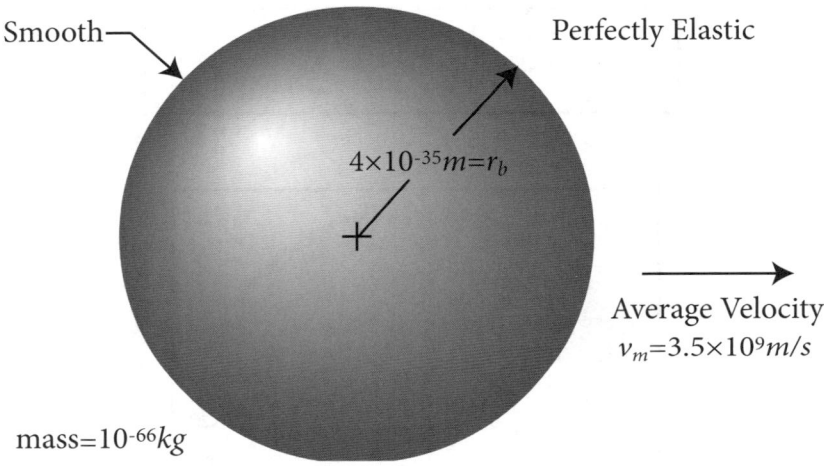

The brutino is the ultimate constituent of the universe. Everything is made of it. There are no smaller particles known.

What is the Smallest Thing in the World? The Brutino

Questions about Brutinos

1. Why were they made spherical?
  > If not spherical they couldn't be smooth and they would hit on edges and, probably, slow down with impacts.

2. Why were they made the same size?
  > Nobody knows.

3. Why were they made with the same mass?
  > Nobody knows.

4. What if the mass of every particle were doubled (or halved).
  > We'd never know the difference. All applied forces would be doubled (or halved) and all inertias would be doubled (or halved).

5. What if the average speed were doubled (or halved)?
  > We'd never have a way to discern the difference.

6. What if the spacing were decreased or the radius were increased.
  > Too much of either would prevent the production of matter in the universe – as we shall later prove.

7. How was the design worked out?
  > Nobody knows.

2. Practically All of the Universe Is Simply a Gas of Brutinos - The Remainder Is Neutrinos and They Are Made of Brutinos

In the beginning there were only brutinos in the three-dimensional space which we call the universe. The spheres, brutinos, are not drawn to scale in the following figure. The cube is drawn to indicate that the brutinos are in a three-dimensional space.

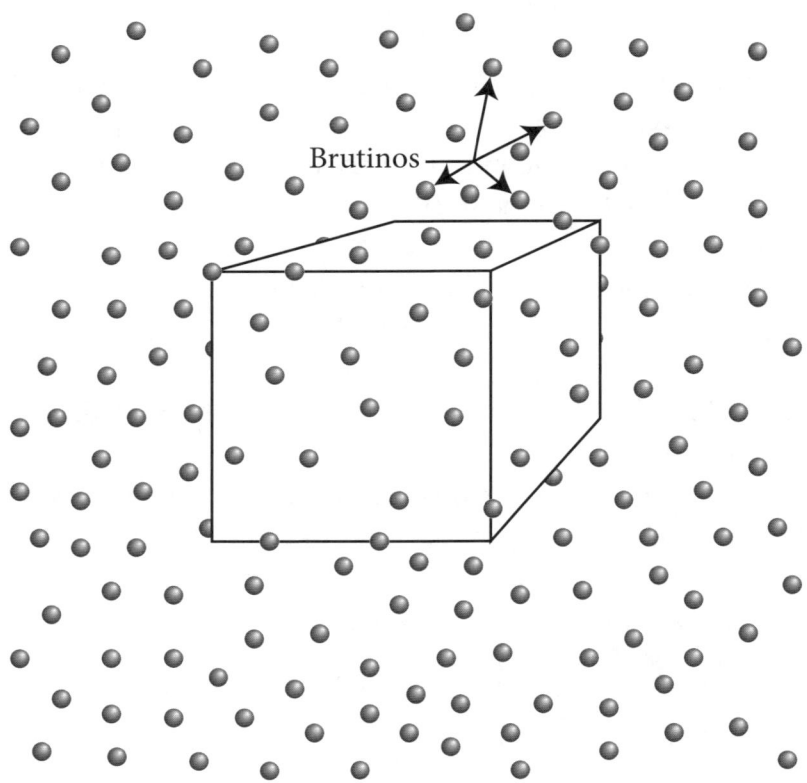

Practically All of the Universe Is Simply a Gas of Brutinos - The Remainder Is Neutrinos and They Are Made of Brutinos

Another picture of the gas is given below. The spheres are represented by *specks*. Their diameters are much smaller than the spacing among them.

The figure below gives values for the parameters defining the brutino gas. The mass of the brutino is 10^{-66} kg. The mass density of the brutino gas is 10^{18} kg/m³. The brutino radius is $r_b = 4 \times 10^{-35}$ m. The average travel distance between collisions is $\ell = 10^{-16}$ meters. The distance ℓ is the mean free path. The particle number density $\eta = 10^{83}$ brutinos per cubic meter. The average distance between brutinos, $s = 10^{-28}$ meters. Notice that ℓ is much greater than s. If the gas particle radius were ten times smaller then the mean free path would be 100 times larger.

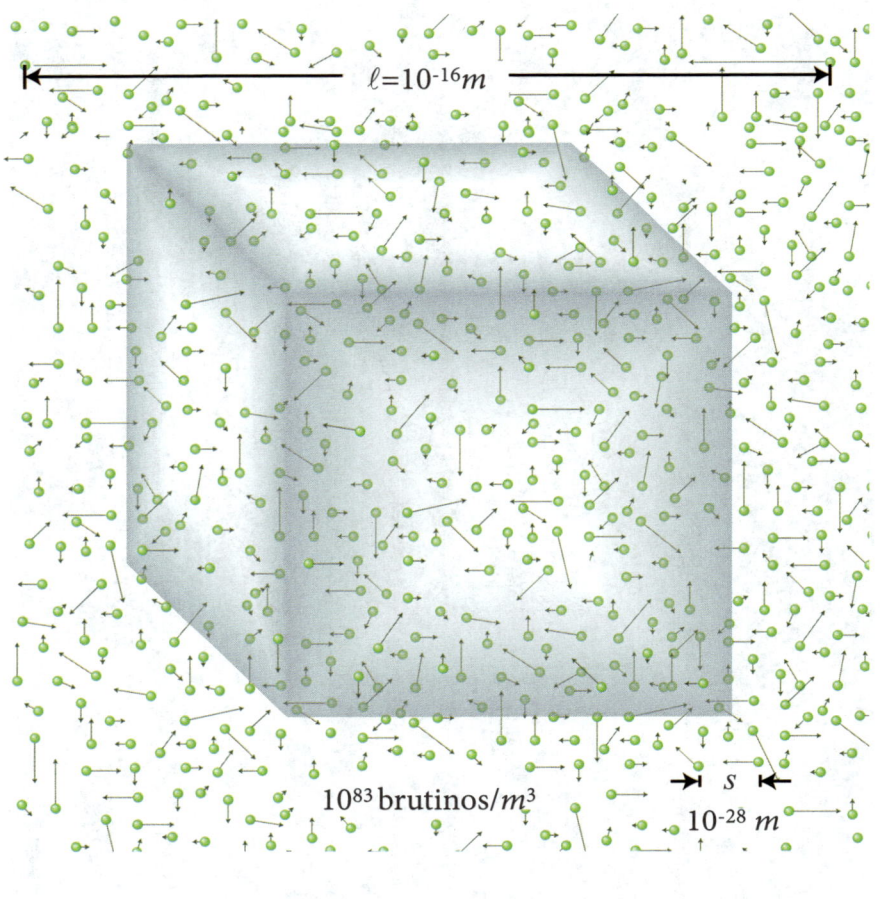

Practically All of the Universe Is Simply a Gas of Brutinos - The Remainder Is Neutrinos and They Are Made of Brutinos

The Brutino Gas (also known as the ether)

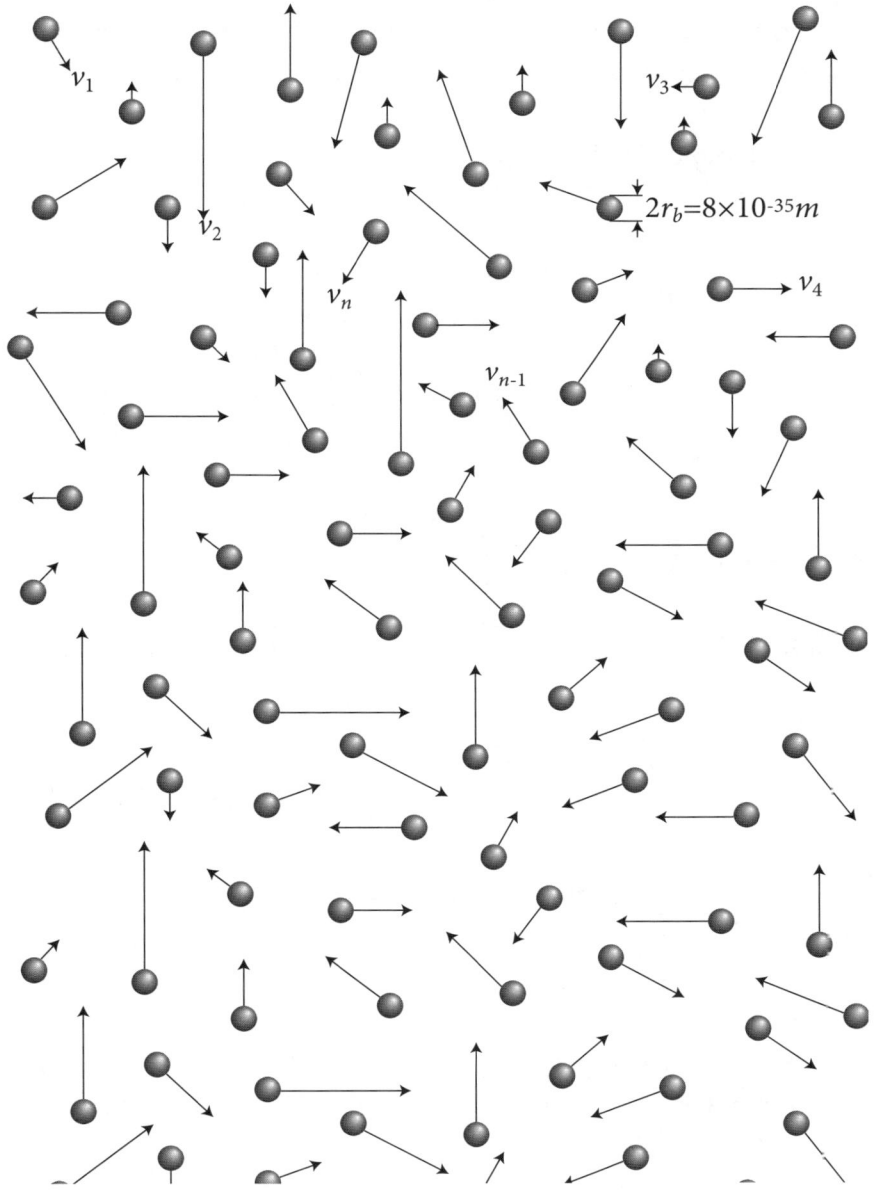

Only one part of space in 10^{20} consists of mass and the rest is a vacuum. The content of the universe is just a rare gas. The brutino gas may be the recently discovered Higgs field and the brutino, or an assembly of brutinos, may be the Higgs (or God) particle.

The average speed of brutinos v_m is given by

$$v_m = \frac{v_1 + v_2 + v_3 + ... + v_n}{n}$$

$$= 3.5 \times 10^9 \text{ meters/second}$$

v_m is the mean speed.

We also will need the average of the squares of the velocities, i.e., the mean square speed.

$$v_r^2 = \frac{v_1^2 + v_2^2 + v_3^2 + ... + v_n^2}{n}$$

$$= 1.5 \times 10^{19} \text{ meters}^2/\text{second}^2$$

v_r is the root mean square speed (rms speed) where velocities are measured relative to frame moving with the gas or relative to a rest frame if the gas is not flowing.

$$v_r = 3.8 \times 10^9 \text{ meters/second}$$

Practically All of the Universe Is Simply a Gas of Brutinos - The Remainder Is Neutrinos and They Are Made of Brutinos

The Universe

Using the universe radius, 10^{26} m, as the scale for this drawing the dots are much too large and their spacing is much too large. The earth, obviously, is drawn much too large.

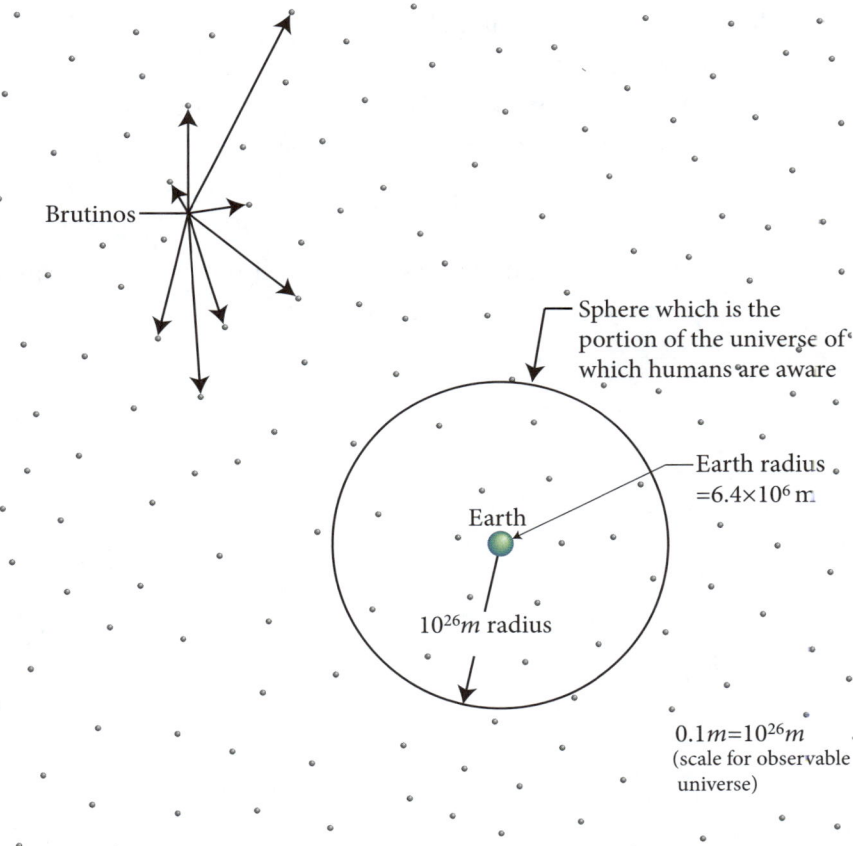

The universe extends indefinitely in three orthogonal (mutually perpendicular) directions.

Questions

1. Why is space three-dimensional, why not two-dimensional, four-dimensional, or even ten-dimensional?
    Unanswerable.

2. How far does space extend?
    Nobody knows.
    We know the universe extends 10^{26} meters but at the present time we don't know how to explore beyond 10^{26} meters from the earth.

3. Some questons are unanswerable. We tell you what we think is answerable and what is not.

Practically All of the Universe Is Simply a Gas of Brutinos - The Remainder Is Neutrinos and They Are Made of Brutinos

The brutinos translate and they collide. They never disappear and never re-appear. When they collide the interaction occurs in zero time, meaning that the brutinos are *hard*. They still are perfectly elastic.

Collision of brutinos

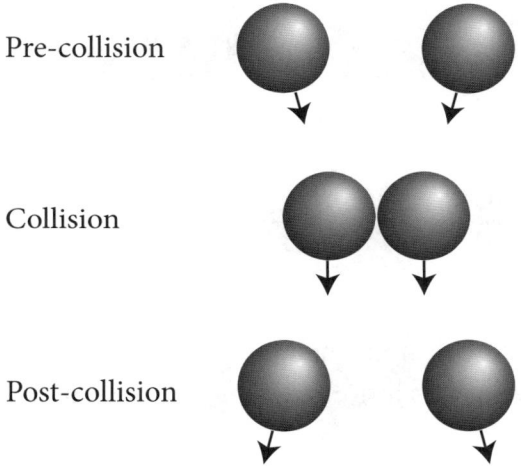

Pre-collision

Collision

Post-collision

Here we show a head-on, equal-opposite velocity collision.

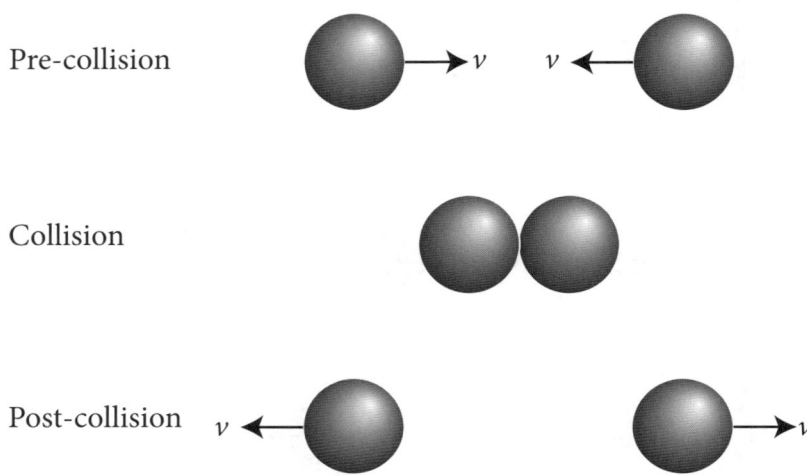

For the head-on collision when the pre-impact velocities are unequal the velocities just interchange particles as a result of colliding.

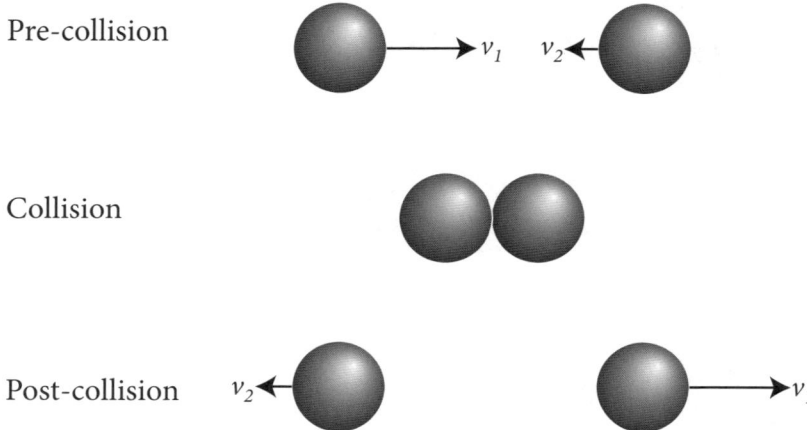

Practically All of the Universe Is Simply a Gas of Brutinos - The Remainder Is Neutrinos and They Are Made of Brutinos

Here we show a general two-dimensional collision. For the general two-dimensional collision, the normal velocity components are interchanged and the transverse velocities are simply unchanged.

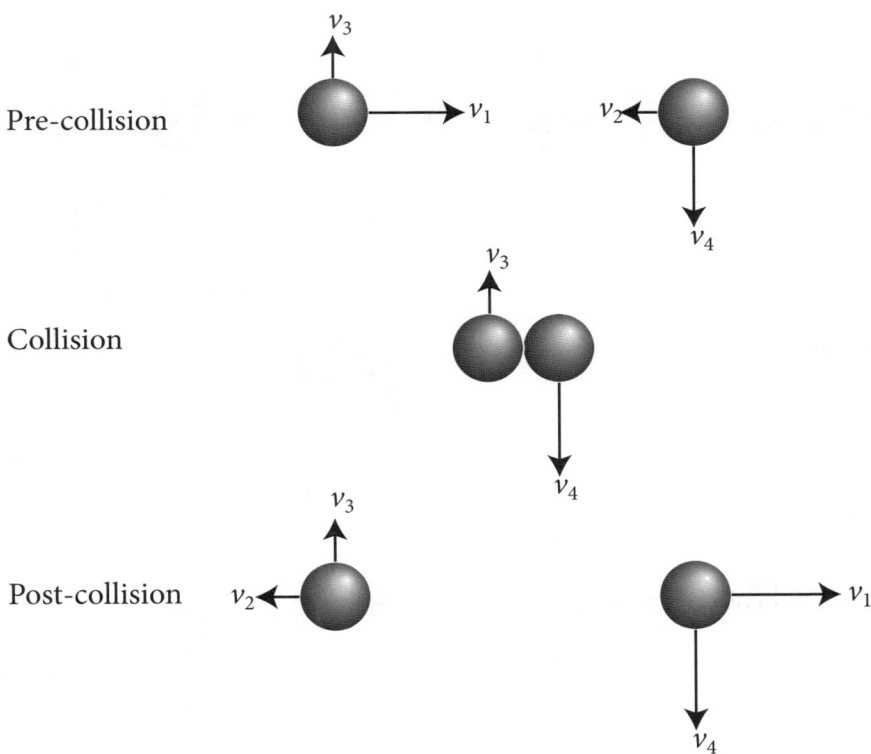

Physics for the Millions

Now we show the general three-dimensional collision.

For the general three-dimensional collision, the normal velocity components are interchanged and the components in the other two directions are unchanged.

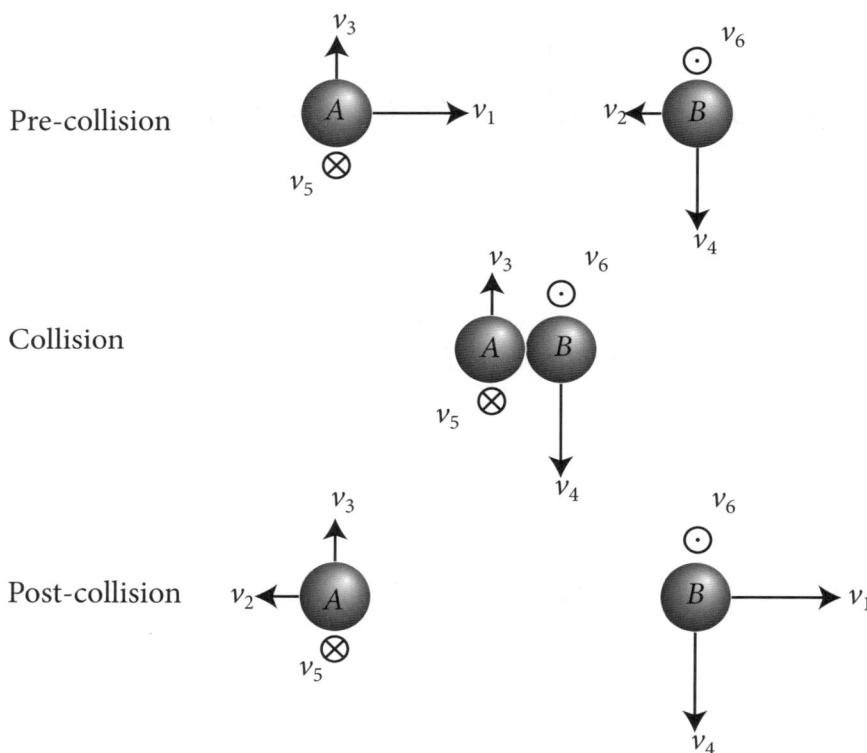

Pre-collision

Collision

Post-collision

Practically All of the Universe Is Simply a Gas of Brutinos - The Remainder Is Neutrinos and They Are Made of Brutinos

The energy of a brutino is its mass times the square of its velocity. The velocity of particle A before collision is $\sqrt{v_1^2 + v_3^2 + v_5^2}$ and of particle B is $\sqrt{v_2^2 + v_4^2 + v_6^2}$. Their energies are $m(v_1^2+v_3^2+v_5^2)$ and $m(v_2^2+v_4^2+v_6^2)$, respectively. After collision their energies are $m(v_2^2+v_3^2+v_5^2)$ and $m(v_1^2+v_4^2+v_6^2)$. We note that the sum of their energies before collision is equal to that after collision, to wit

$$m(v_1^2+v_3^2+v_5^2) + m(v_2^2+v_4^2+v_6^2) = m(v_2^2+v_3^2+v_5^2)+m(v_1^2+v_4^2+v_6^2)$$

Since the total energy of brutinos does not change if there are no collisions and since it does not change when there are collisions, it follows that the energy of brutinos is forever constant.

The Initial Contents of Space

Brutinos are perfectly elastic, smooth spheres that are all the same size and that make up a gas which is spread somewhat uniformly throughout all of space. Being smooth means that their transverse velocities do not change as a result of a collision

Each brutino has a radius of 4×10^{-35} *m* and a mass of 10^{-66} *kg*. The average mass density of the ether gas is 10^{18} *kg/m³*. The brutinos have an average spacing of 10^{-28} *m*. The brutinos have an average velocity of 3.5×10^9 *m/s* – approximately ten times the speed of light. Even though the average spacing is 10^{-28} *m* each brutino travels 1,000,000,000,000; or 10^{12} times the spacing distance, on the average, before impacting another brutino. If the brutino radius were reduced by a factor of 10, the average distance traveled between impacts would be 100 times larger – if the radius were increased by a factor of 10 the average travel distance would decrease by a factor of 100. Even though the brutino gas has a mass density of 10^{18} *kg/m³*, space is mostly vacant. Brutinos occupy, on the average, one hundred billion billionths (1 part in 10^{20}) of space.

Practically All of the Universe Is Simply a Gas of Brutinos - The Remainder Is Neutrinos and They Are Made of Brutinos

There are two significant questions about a theory of physics based on Newtonian particles – i.e., particles which are hard, smooth, perfectly elastic, and with no force fields. The brutinos only interact when they meet and the only force is a large force of repulsion upon contact. First and foremost, how can such particles form long-lived condensations? Second, how can matter move in such a dense background gas? The density of the background is some 10^{14} (i.e., 100 trillion) times the density of lead. The answers to both these questions come from condensations which are inside a volume of space whose diameter is the mean free path and harnessing the translatory motion of neutrinos into circular paths that do not translate but which can be changed to two-dimensional spiral paths that result in translation.

Before we discuss the condensations which produce the neutrino, we need to discuss the way brutino speeds are distributed. We need to show what happens when background (ether) particles have been aligned and then are completely condensed.

Let us present an example of a velocity distribution which changes with time. Consider a perfectly elastic cubic box of gas of N identical elastic particles of mass m uniformly distributed throughout the box. All have exactly the same speed v. Further, initially 1/6N move to the right, 1/6N move to the left, 1/6N move up, 1/6N move down, 1/6N move toward the back, and 1/6N move toward the front. Initially the mean (average) velocity is

$$\bar{v} = \frac{v_1 + v_2 + v_3 + \ldots + v_N}{N} = \frac{v + v + v + \ldots + v}{N} = \frac{N(v)}{N} =$$

$$= v$$

The initial mean square velocity is

$$\bar{v_r^2} = \frac{v_1^2 + v_2^2 + v_3^2 + \ldots + v_N^2}{N} = \frac{v^2 + v^2 + v^2 + \ldots + v^2}{N} = \frac{N(v^2)}{N} =$$

$$= v^2$$

The rms velocity is

$$v_r = \sqrt{v^2} = v$$

Given the initial conditions, where every particle has the same speed and all velocities parallel to various sides of the box, they begin to collide and change their directions. However, with the assumptions specified, their speeds do not change. We show how the directions change and how the speeds do not change with the following illustration. Consider two particles approaching each other with equal and opposite velocities of 10 *m/s* and let their contact be such that the line of centers makes an angle of 30° with the velocity vectors.

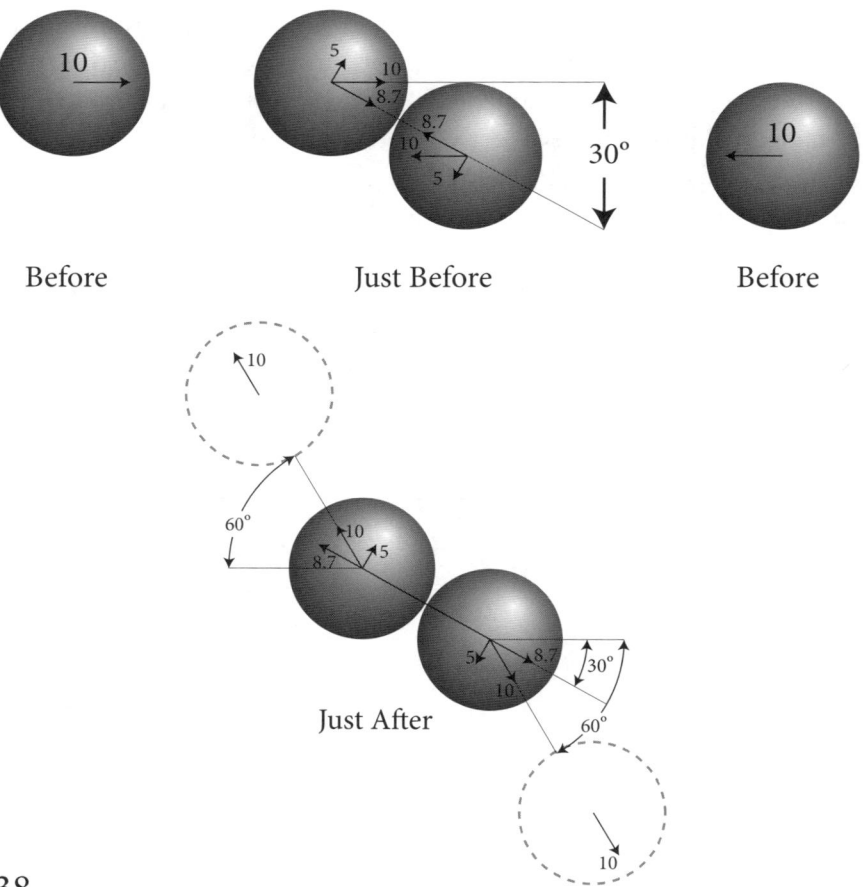

Before   Just Before   Before

Just After

38

As a result of repeated collisions, velocities in all different directions will be obtained but the speeds remain the same.

There are a number of ways to obtain varying speeds. If the particles had a slight variation in mass various speeds would be obtained. Also, three-particle collisions could produce speed variations — but that would require a non-zero interaction time. We assume that the speeds change as a result of collisions.

After a large number of collisions, not only will the particles have all different directions but also the number of particles having velocities between equal increments of velocity will be as illustrated below. N_i is the number of particles having velocities falling in the ith velocity increment. The particle velocity directions will be equally likely over 4π sterradians.

Given the initial conditions where every particle has the same velocity, the particles begin to collide with each other. As a result, some particles will move faster than v_{mp} and some slower than v_{mp}.

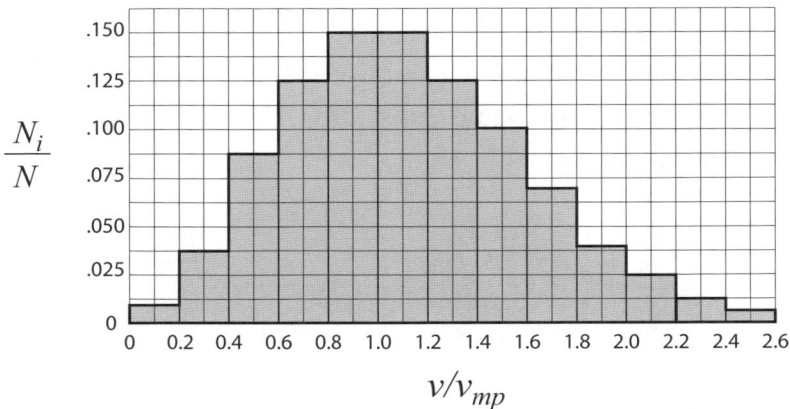

This above plot shows the portion of particles having velocities that are separated by the velocity increment $(1/5)(v/v_{mp})$. For example, for the velocity increment $.6v_{mp}$ to $.8v_{mp}$ the portion is .125 or 12.5

percent. Twelve and a half percent of the particle velocities will be between $.6v_{mp}$ and $.8v_{mp}$.

If N is a very large number and small increments of velocity are used, the plot similar to the above approaches being a smooth plot, see the figure.

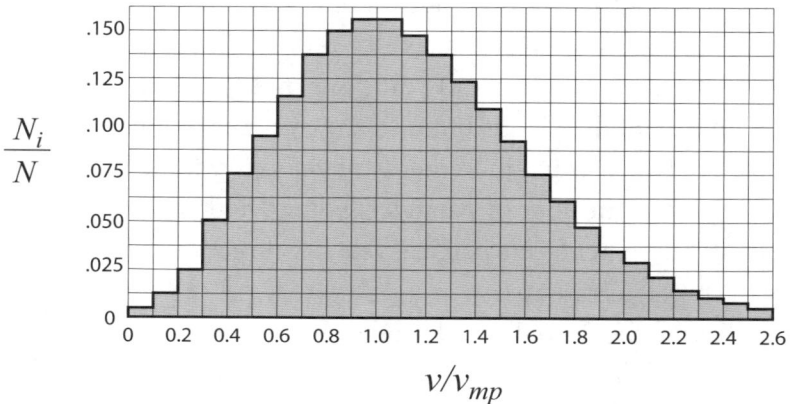

If this process is continued by letting N increase indefinitely, and with the velocity increments decreasing indefinitely provided that the number of particles in each increment remains large, then a smooth curve will be obtained and we let N_i/N become $f(v/v_{mp})$. In this expression v_{mp} is the value of abscissa where f is a maximum. We show the resulting curve.

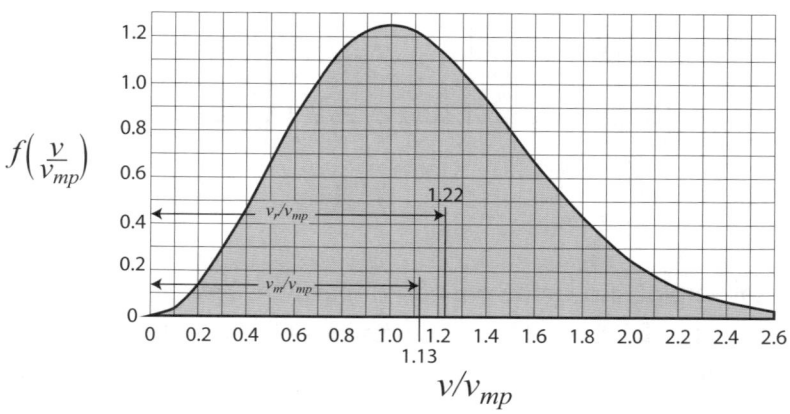

Practically All of the Universe Is Simply a Gas of Brutinos - The Remainder Is Neutrinos and They Are Made of Brutinos

This plot is *normalized* so that the area under the plot is unity. Now this plot will give the percentage of the velocities falling between 0.2 and 0.3, for example. In this case the percentage is 100(0.023)=2.3. Thus 2.3 hundredths of the velocities in the elastic box will occur between values of v/v_{mp} from 0.2 to 0.3. But, what is the value of v_{mp} for the box we initially discussed?

There are two other velocities which we need to identify on this plot. They are the mean velocity and the rms velocity. The mean velocity is 1.13 times the most probable velocity, and the rms velocity is $\sqrt{3\pi/8}$ =1.08 times the mean velocity or 1.22 times the most probable velocity. The value of v_{mp} is 0.855 times v_m and 0.820 times v_r. We show these two velocities on the plot.

This plot of the probability of a given velocity was obtained by theoretical arguments initially by James Clerk-Maxwell and was later obtained more rigorously by Ludwig Boltzmann. It is known as the Maxwell-Boltzmann distribution function. A large number of experiments with actual rare gases have verified this theoretical distribution. The ratio of the rms speed to the mean speed is obtained from the theory and its value is $v_r/v_m = \sqrt{3\pi/8}$.

Let us now return to the elastic box with the elastic particles whose speeds can change. All particles were initially translating at the velocity v so that the mean velocity was v and the rms velocity was v. After sitting for a while the particles began colliding and no energy was added to the box so that the rms velocity remained at v. On the other hand the mean velocity was reduced to $v_m = v_r/\sqrt{3\pi/8}$. Thus, the mean velocity decreased by almost 8 percent.

We will later show that nature has evolved a process of taking randomly moving particles with a mean velocity v_m, aligning them, keeping their energy the same, and increasing their mean velocity by the factor 1.085. This process occurs in the neutrino and it is the source of everything observed in the universe.

Incidentally, the mean speed will always be less than the rms

41

speed unless the particles are all moving at the same speed. This is illustrated by the following example. Let two particles be moving parallel to each other in the same direction. Let one have a velocity of 9 meters/sec and the other 11 meters/sec. Their average speed is

$$v_m = \frac{9+11}{2} = \frac{20}{2} = 10 m/s$$

Their rms speed is

$$v_r = \sqrt{\frac{9^2+11^2}{2}} = \sqrt{\frac{81+121}{2}} = \sqrt{101} = 10.05 m/s$$

which is one half percent higher than the mean speed.

One other phenomenon we will need to establish is the mechanism by which a neutrino develops a thrust.[1] Consider a number of brutinos which were taken from the background and aligned to all move in the same direction without changing their speeds, see the figure.

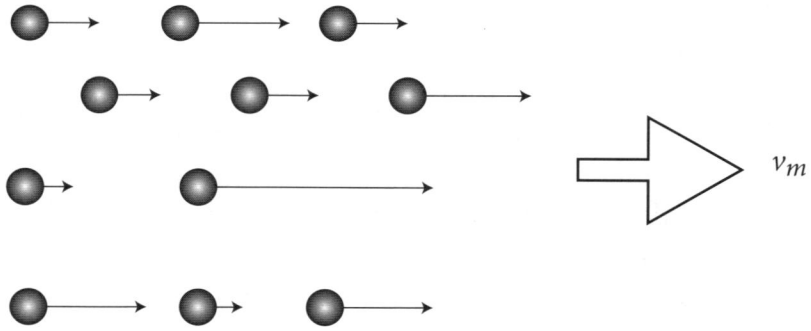

Their flow speed is v_m since in the background their average speed was v_m. Their rms (absolute) speed also is v_r, the background rms speed. We use absolute here to refer to the square root of the sum of the actual particle speeds rather than the usual definition of the rms

---

[1] We think of a thrust as a force which moves and thus does work.

speed which speeds are the velocities relative to the flowing particles.

In order to understand the propulsion system of the neutrino, consider first a spherical system at rest which is impacted by particles coming in spherically symmetrical and being scattered back along a diameter as shown here.

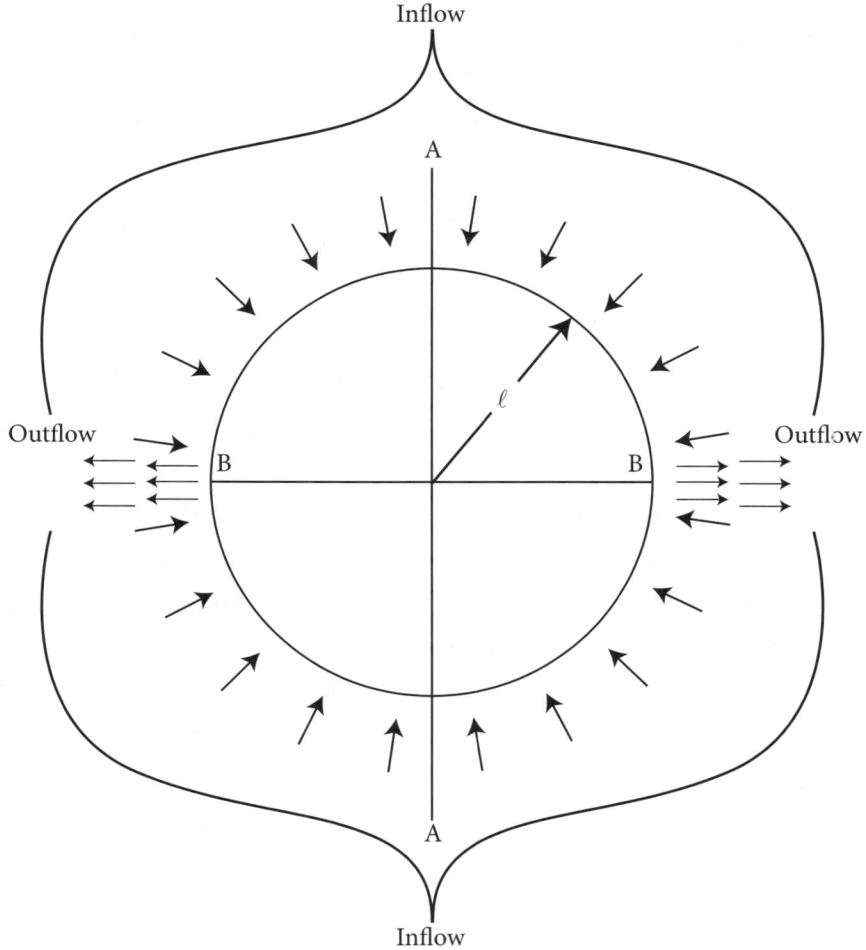

If the particles enter spherically symmetrical and exit out the two cylindrical sections we then can calculate the total force on the plane through AA perpendicular to the paper. The mass inflow rate impinging on plane AA from the right is $\rho Av = \rho(2\pi\ell^2)v$, where ρ is the mass density, ℓ is the radius of the sphere, and v is the flow velocity. The force applied to the plane AA by particles impinging from the right is the component of momentum per unit time impinging plus the momentum per unit time rebounding. The average momentum impinging in the direction BB is the centroid portion of a sphere (0.5 times the radius) times the total impacting, and the rebounding is double this. Thus the force on plane AA can be approximated by

$$F_{AA} = 1.5 \, \rho(2\pi\ell^2)v_m \times 0.5 v_m$$
$$= 1.5 \times 4.23 \times 10^{17} [2\pi(2.35 \times 10^{-16})^2](3.51 \times 10^9)^2$$
$$= 2.71 \times 10^6 \text{ Newtons}$$

The force is actually about 17% larger than this since the plane AA has a velocity perpendicular to AA to the right. This is a very large force!

If the neutrino did not move there would be this force impinging on the right side of plane AA and an equal force impinging on the left side of plane AA. Of course, no work would be done. However, on one end of the core, motion is occurring due to the manner in which the neutrino was made. It was made with a solid core at one end and that end is propelled by the following mechanism.

Consider a stream of particles aligned parallel to each other moving in the same direction, as shown on page 42. Let the particles be brutinos taken randomly from the background gas and aligned without changing their speeds. Their flow velocity will be v_m since that was the mean speed in the background. Their energy would be Nmv_r^2 for the same reason. Now, let these particles be impacted from the sides without changing their energies and let the particles continue flowing from plane AA until all the particles all are touching each other – i.e., until they are packed solid. Also, assume during this process that the energy of the stream of particles did not change. Since the particles are all moving at one value of velocity the velocity must be v_r, the background rms velocity. Thus, their flow velocity initially was v_m and now it is $v_r = \sqrt{3\pi/8}\, v_m = 1.085\, v_m$, then the flow velocity increased 8.5 percent. This is the mechanism which propels the neutrino.

The forward moving stream experiences a force of over a million Newtons and the forward moving particles flow at velocity v_r while the rearward flowing particles flow at velocity v_m. The net result is that the neutrino translates at the velocity $v_r - v_m$ and the neutrino speed is the speed of light. Thus, the speed of light c is

$$c = v_r - v_m$$

3. Neutrinos Are Made of Brutinos

Neutrinos were the first entities appearing in the universe beyond the gas of brutinos (which we call the ether). The brutinos condense in small regions of space to produce neutrinos.

The neutrino diameter is approximately the same as the average distance traveled by a brutino from one collision to the next. The neutrino has a semi-solid near cylindrical core whose length is approximately equal the neutrino diameter and whose radius may be $1/1,000,000,000^{th}$ (one billionth) the diameter of the neutrino.

The free neutrino translates with a direction parallel to the longitudinal cylindrical axis.

The neutrino is somewhat like an atmospheric tornado but is much more stable and it translates somewhat faster than a tornado (about 8% the mean speed of the ether gas) when comparing translatory speeds with gas particle speeds.

The neutrinos twist as they traslate.

Neutrinos Have Angular Momentum
Neutrinos Twist as they Translate

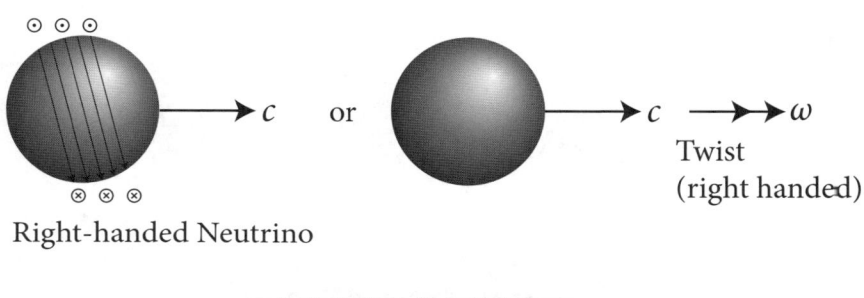

Right-handed Neutrino

Twist (right handed)

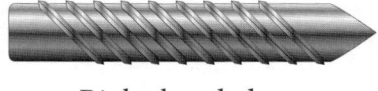

Right-handed screw

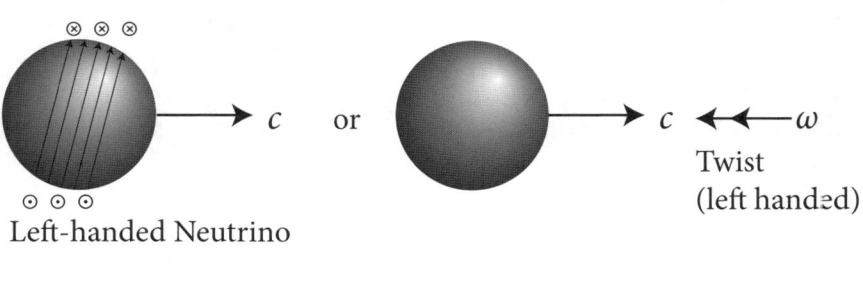

Left-handed Neutrino

Twist (left handed)

Left-handed screw

Formation of Neutrinos

The neutrino formed in the ether gas is much like a tornado formed in air. Due to the random motion of the ether gas particles winds can build up. These ether winds, just as winds of the atmosphere, can spawn the *tornado-like* neutrinos.

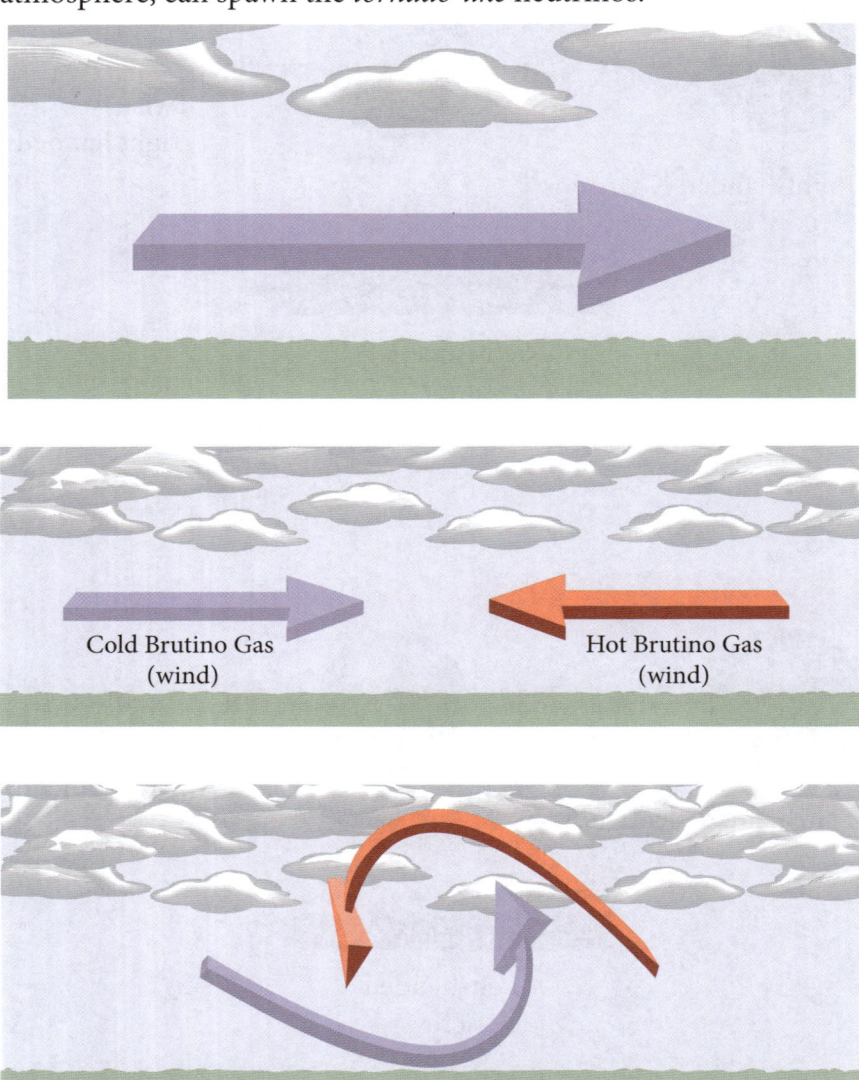

Cold Brutino Gas (wind)

Hot Brutino Gas (wind)

Formation of a neutrino (continued)

Formation of a Neutrino (continued)

The neutrino is expelled from the billowing clouds.

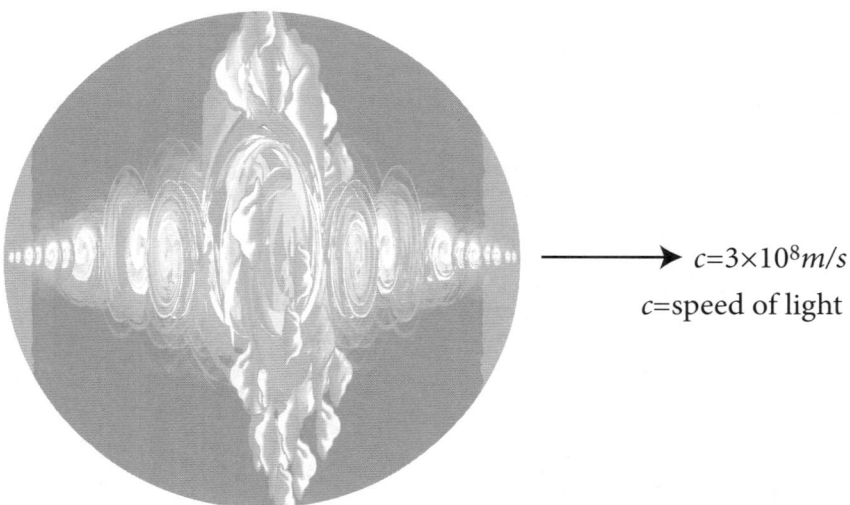

$c = 3 \times 10^8 m/s$
c = speed of light

The Completed Neutrino

Cloud picture of neutrino

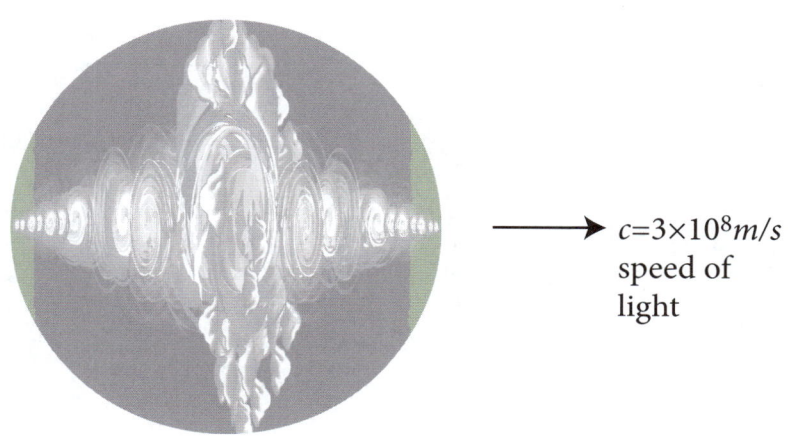

$c = 3 \times 10^8 m/s$ speed of light

Particle picture of neutrino

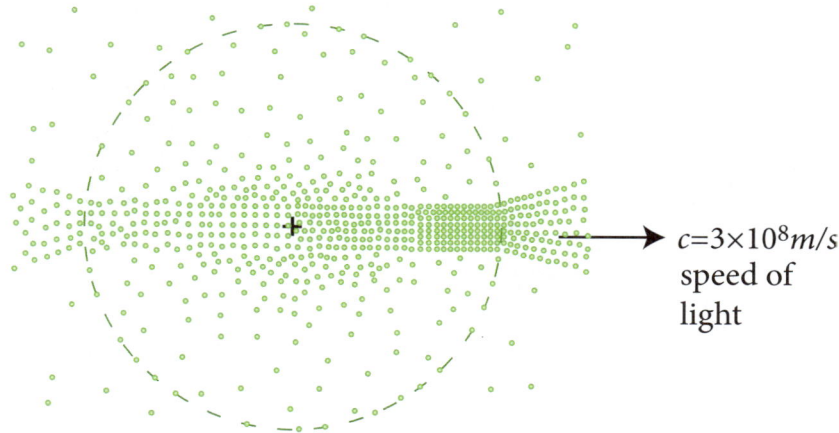

$c = 3 \times 10^8 m/s$ speed of light

Particles in a neutrino

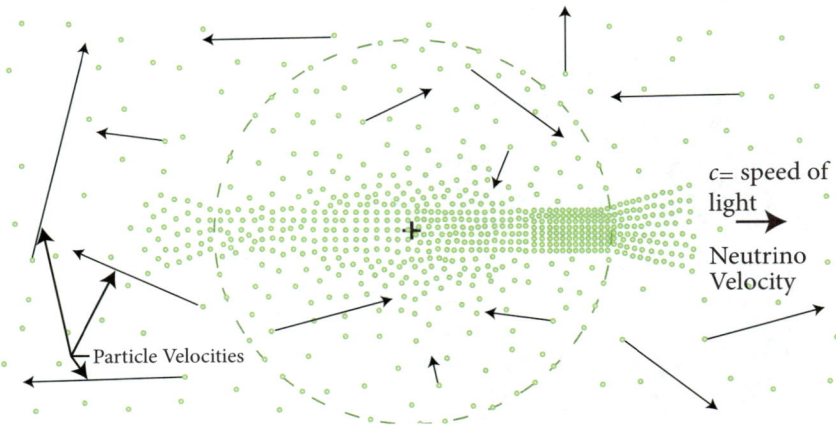

Flows in a neutrino

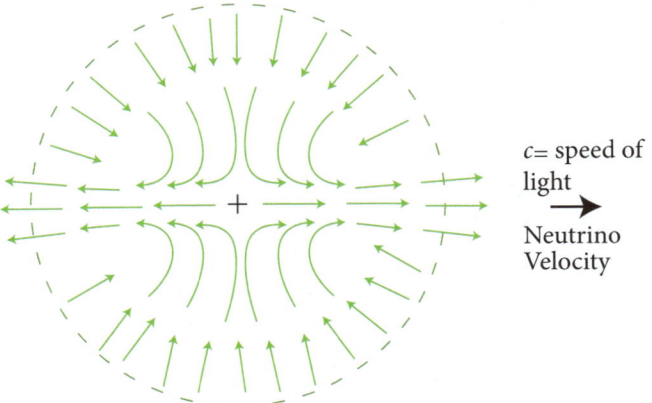

Structure of the Neturino

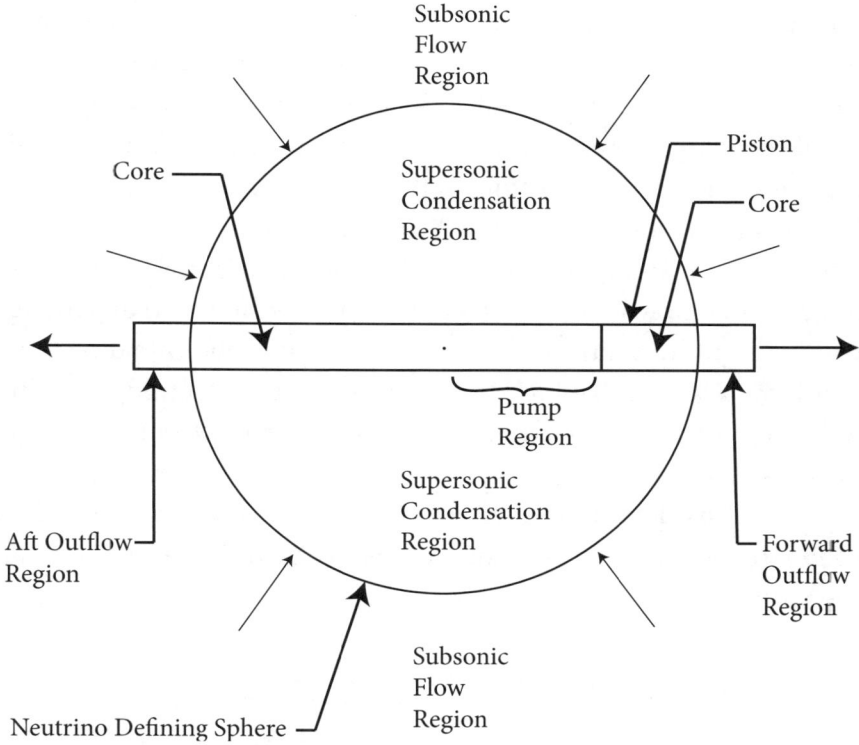

The *neutrino defining sphere* is the surface separating the subsonic flow region from the supersonic (free-molecular-flow) region.

Flow Requirement in the Neutrino

To form a neutrino it is necessary to form a solid core of just the correct size. If the core is too large then particles coming in to the center from one direction will meet particles coming from the other direction which will disrupt the flow. If the core is too small the *pumping* rate will be so low that there will not be enough particles to form a solid piston. If the core is *just right* between these two situations, then the flow will *go critical* and form the neutrino. The size of the pump will be such that the particle flow rate will be large enough to sustain the solid pump and low enough that particles coming into the neutrino-defining sphere from opposite directions will have a low probability of impacting particles coming in. The following figures illustrate the *too small, just right, and too large* piston sizes.

Let us determine the size of the sonic sphere to produce a solid stream with a cross section of one brutino diameter. Let ℓ be the radius of the sonic sphere. Now

$$\rho_o (2\pi \ell^2)(0.7 v_m) = \frac{m_b 4\pi r_b^2}{(4/3)\pi r_b^3} v_m$$

or

$$\ell = \left[\frac{m_b 4\pi r_b^2}{(4/3)\pi r_b^3 \rho_o 2\pi (0.7)} \right]^{1/2} = \left[\frac{2.89 \times 10^{-66} \times 3}{(1.4\pi)(4.05 \times 10^{-35})(4.23 \times 10^{17})} \right]^{1/2}$$

$$= (0.115 \times 10^{-48})^{1/2} = 0.339 \times 10^{-24} = 3.39 \times 10^{-25} m.$$

This value of the sonic radius is the absolute minimum value for the size of a neutrino. It is expected that a much larger solid stream of particles, rather than one which has a cross section of only one diameter wide, would be required for stability. The optimal value of ℓ is believed to be nine orders of magnitude larger than this.

Flow Requirements in the Neutrino (continued)

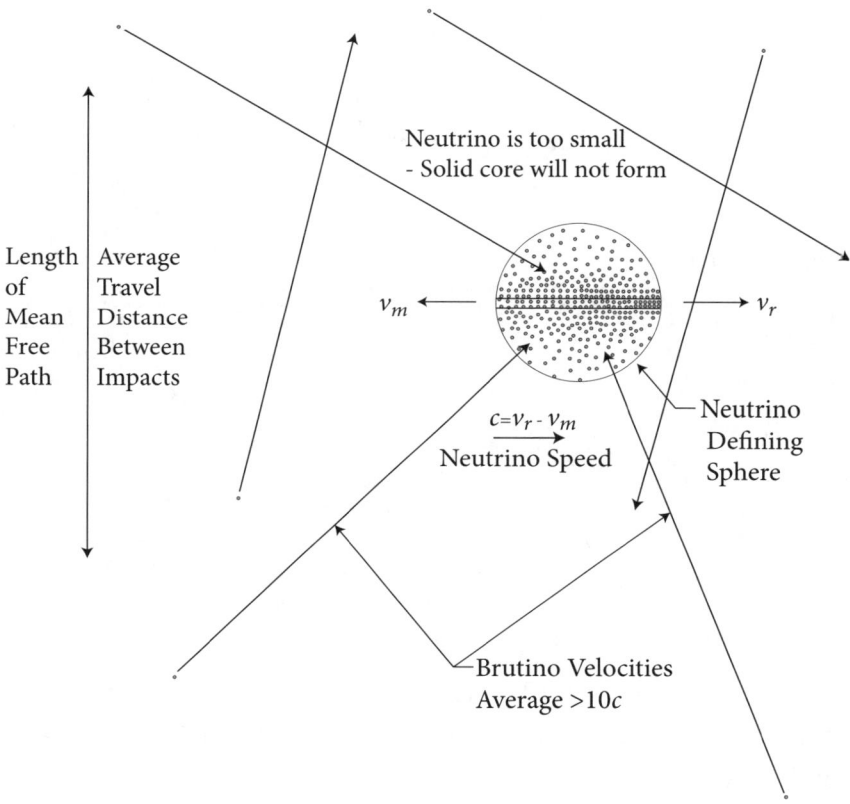

Mean free path is longer than neutrino diameter. Solid core will not be maintained. Neutrino will not survive.

Flow Requirements in the Neutrino (continued)

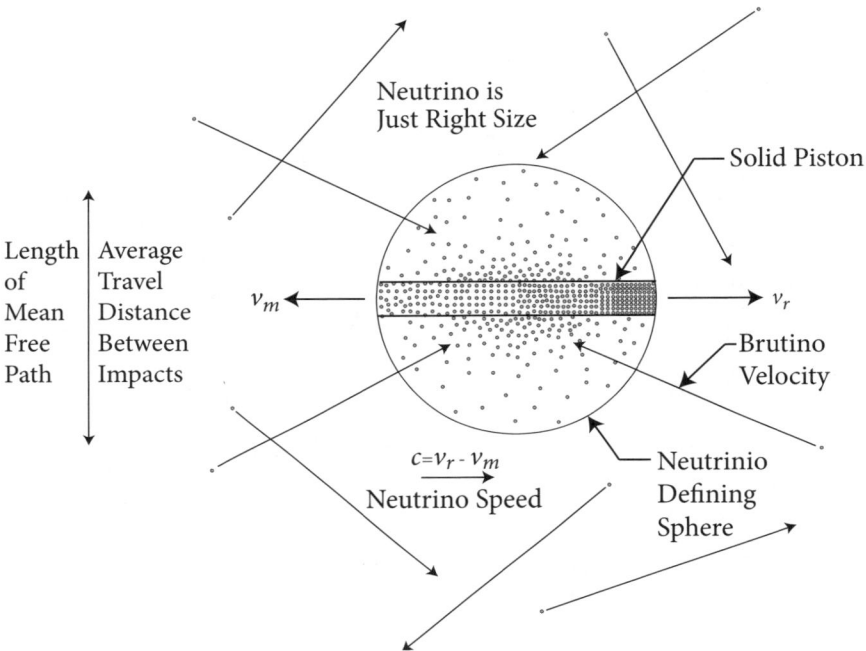

Mean free path is about the same as the neutrino diameter. Solid core will be established. Impacting brutinos will not impede condensation. Neutrino will survive.

Flow Requirements in the Neutrino (continued)

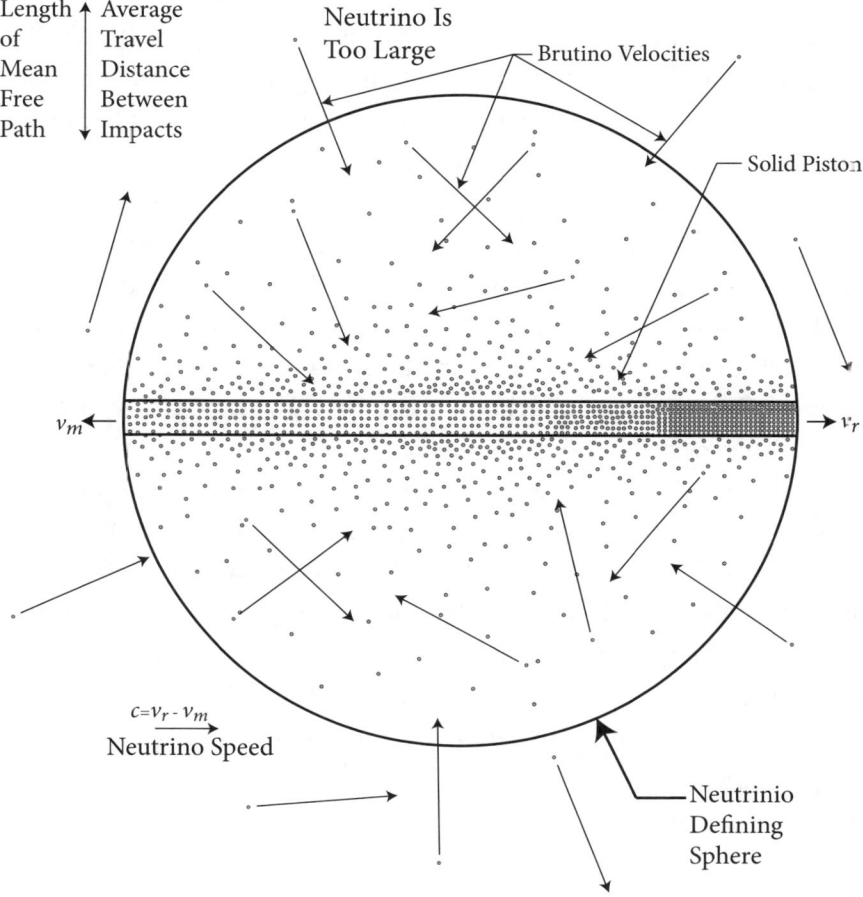

Mean free path is short compared to the neutrino diameter. Solid core could be established but flow into core is impeded by impacts since mean free path is short. Neutrino will not survive.

There are two secrets which nature hid from mankind for the many years during which scientists searched for such things as condensations of a gas of inert elastic particles. These secrets are the *pump* mechanism and that the condensation had to occur within a nuclear-sized volume and that the nuclear size is the size of the mean free path (i.e., a dimension on the order of $10^{-16}m$).

Either end of the core can get solidified but only one end does since when the solidification occurs, the solid core begins moving in the direction of the solid core and takes the whole assembly with it. This motion prevents the other end of the core from solidifying. This solidification moves forward and produces a lowered pressure region behind it which sucks in particles. The amount of suction depends upon the size of the solid core.

When the core size is optimal there will be enough particles flowing in to form the semi-solid region of the core just aft of the solid core. Then the flow rate down to the core will be small so that the character of the flow will be *free-molecular-flow*. If the core is too large the in-flow will be more like a *continuum of gas* and particles coming from all directions will interrupt the inflow. This occurs when the mean free path length is small compared to the sonic-sphere radius.

The neutrino is the *engine* which provides the energy to drive the universe. The way this works is that particles coming into the neutrino generally from the forward hemisphere impact those coming into the neutrino from the aft hemisphere as illustrated below. However, the forward half of the core becomes solidified and forms the solid piston. The solidification makes a region which moves forward so that the impacting force moves, thus becoming a thrust. The thrust does work and propels the neutrino forward. The following figure illustrates the impacts and flows.

Vectors showing the impacts driving the neutrino

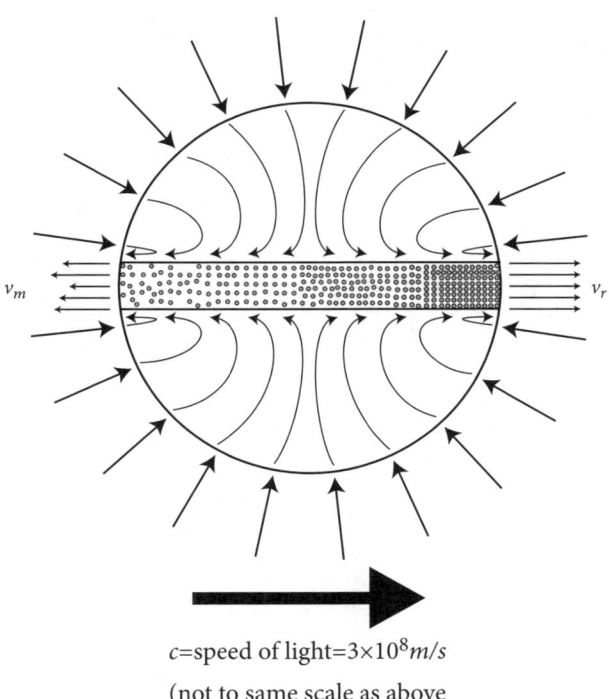

c=speed of light=$3 \times 10^8 m/s$
(not to same scale as above arrows)

Physics for the Millions

The figure shows the magnitude of inflow velocities and the two outflow velocities.

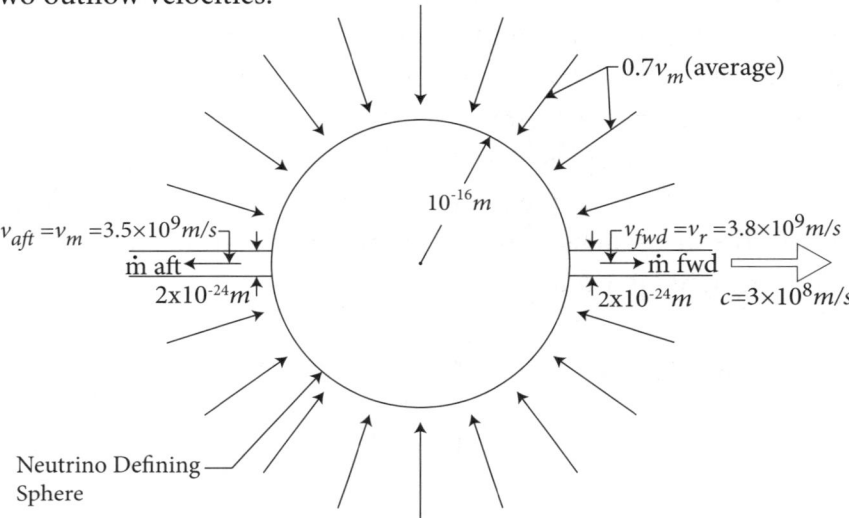

The following figure shows the core region and flow velocities.

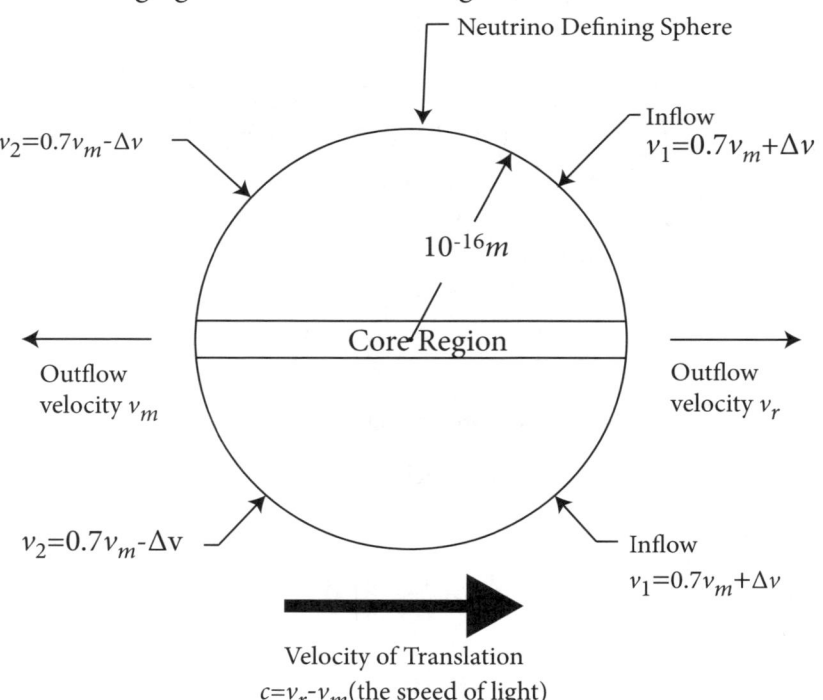

Neutrinos Are Made of Brutinos

There are three regions of flow of the neutrino
1. The subsonic flow outside the defining sphere.
2. The supersonic (free molecular) flow from the defining sphere surface to the core.
3. The core structure which consists of a pump and a piston.

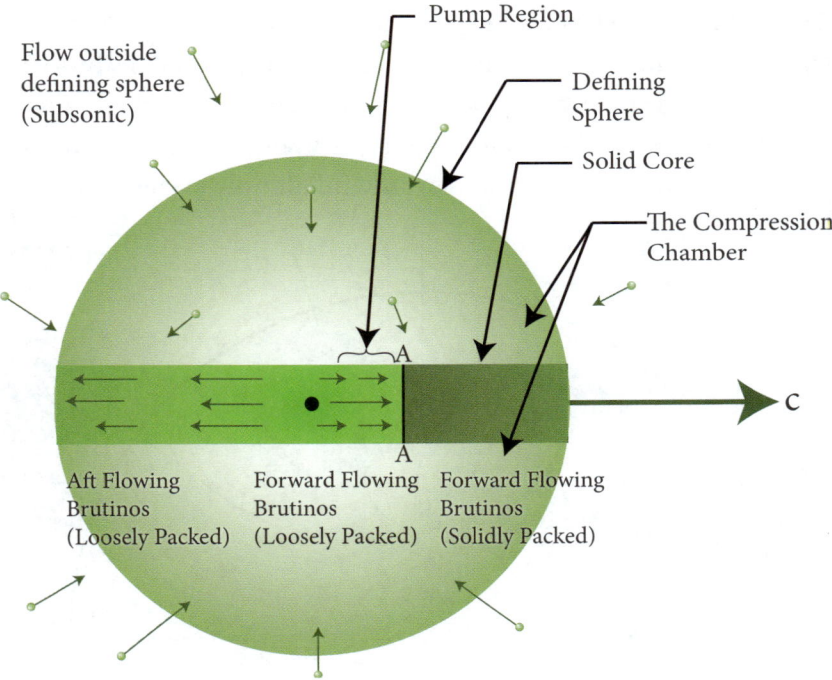

The figure above shows a schematic which emphasizes the core structure.

The flow inside the neutrino is actually more complicated than we have illustrated in the previous figures since the particles circulate about the neutrino cylindrical axis as they come in from all directions and exit from two *holes*, i.e., the forward end of the core and the aft end of the core. These two exits are each similar to the drain from a bath tub.

What occurs to cause circulation around a sink is that the pressure on a plane perpendicular to the flow is higher than that transverse to the flow so the fluid begins flowing laterally to the original flow. The flow thus begins circulating around the diameter which is parallel to the displacement. The figure below shows the circulatory flow as well as the forward and aft flow at the center.

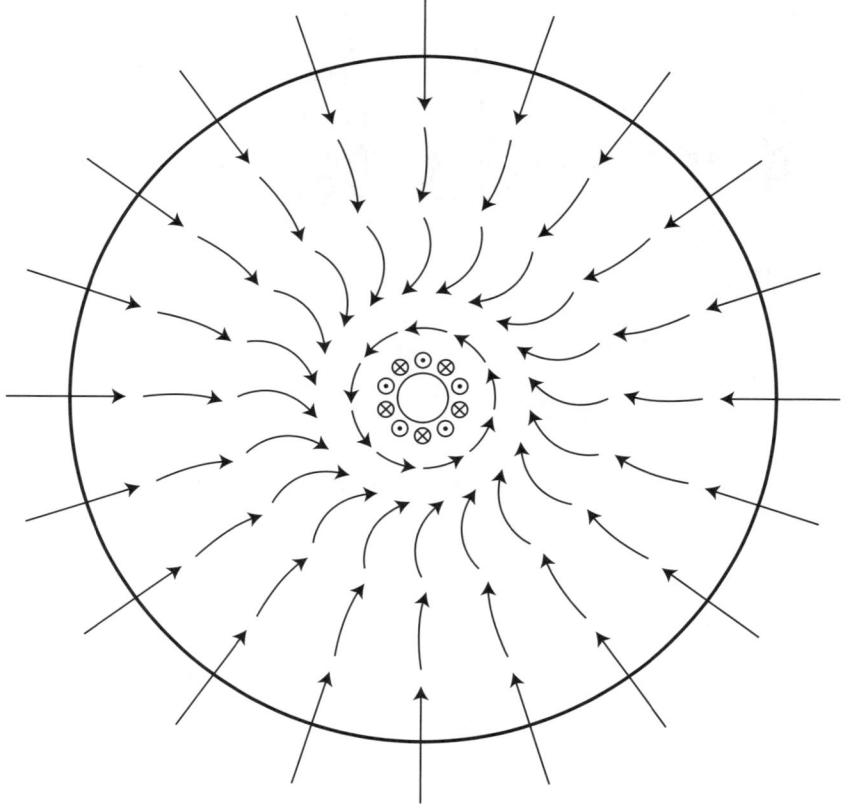

⊙ Vector for forward flow
⊗ Vector for aft flow

We have just shown how the universe began as a space containing a gas of elastic spherical particles which translate and collide. The random motion of the particles forms *winds* and some winds spawn ether gas tornadoes which have long *lives*. These tornadoes are neutrinos and they make the contents of the universe with which we are familiar.

The above explanation of the beginning of the universe consisting of a three-dimensional space filled with a gas of spherical particles is contrasted with the *big bang* model. In the big bang model space begins with infinitesimal dimensions and infinite mass density. At the beginning of the big bang the dimensions begin expanding and matter and other contents begin expanding and taking form. *Taking form* is somewhat similar to the growth of an organism starting with a fertilized cell. In the ten billion (or so) years from the start of the universe we see what we have today. And today, according to the big bang theory, we are still expanding.

With the kinetic particle theory of the universe we see that the universe actually began filled with a gas of perfectly elastic spherical particles which are all alike. Next we see that winds are formed in the gas as a result of the random motion of the particles. These winds spawn small, long lasting tornadoes which are the neutrinos which populate the universe. In the following chapter we will show how elementary matter particles are formed. In later chapters we show how stars are formed and how our planet was formed.

4. The Proton and Electron are Made of Neutrinos

The proton is produced by a right handed neutrino having a mass of $1.6 \times 10^{-27} kg$ getting knocked into a circular orbit. The figure shows a proton-sized neutrino in the upper left of the figure translating to the right in a straight line at the speed of light c. The massive neutrino gets impacted almost simultaneously by several other large neutrinos. We show these neutrinos on the lower right of the proton-sized neutrino.

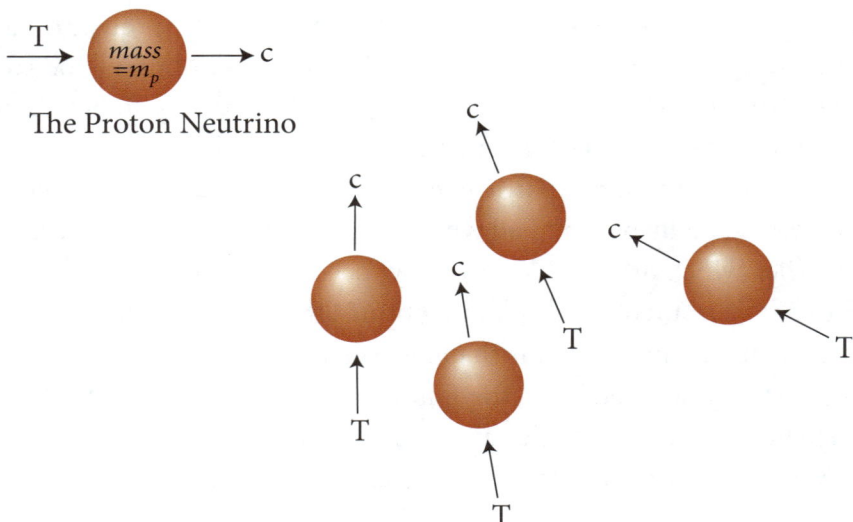

The Proton Neutrino

Massive Neutrino Impacted by Several Neutrinos Near Simultaneously

The Proton and Electron are Made of Neutrinos

Proton-Sized Neutrino Getting Knocked Toward a Circular Orbit

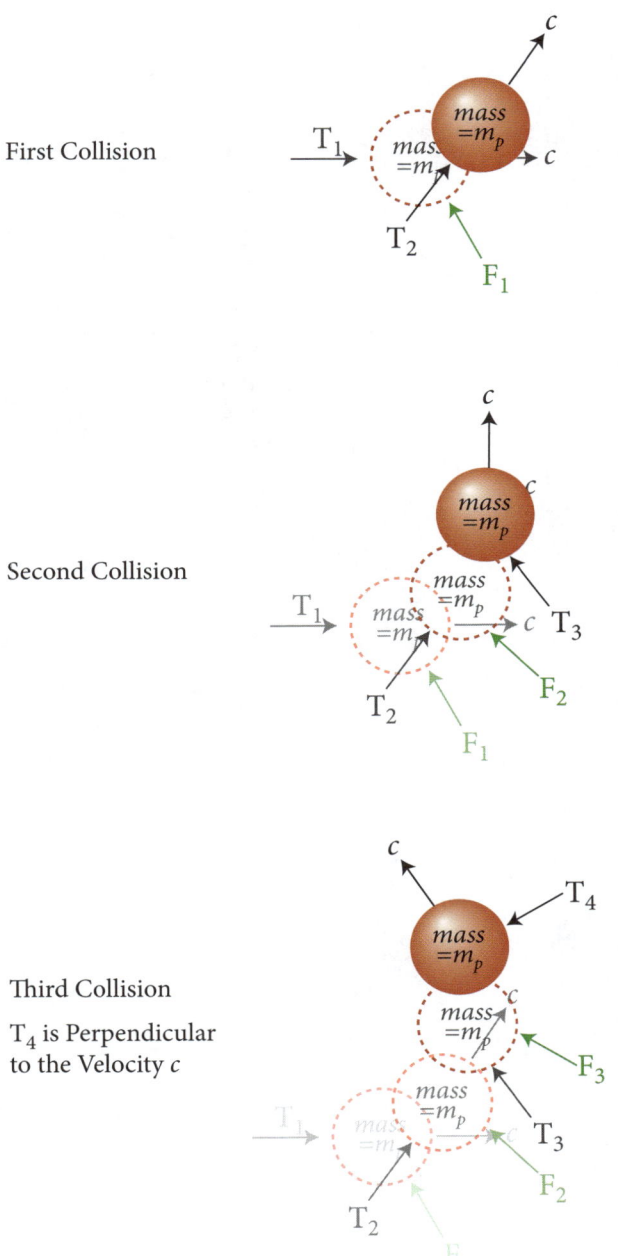

First Collision

Second Collision

Third Collision

T_4 is Perpendicular to the Velocity c

Physics for the Millions

Circular Orbit Is Obtained

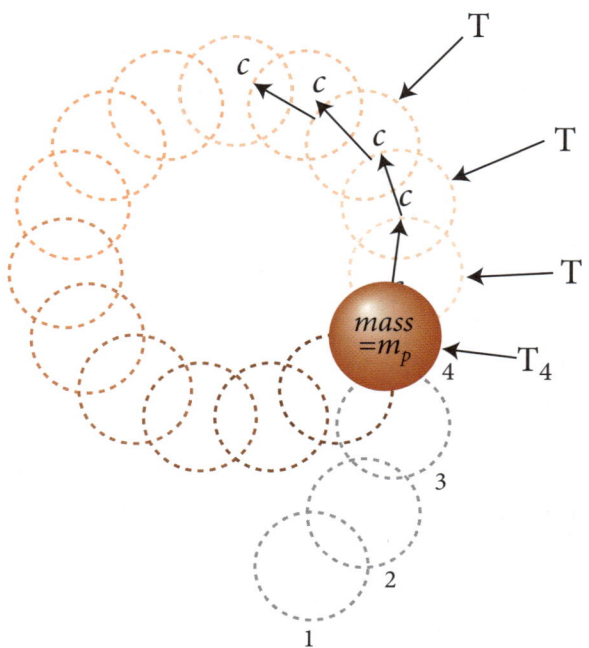

The Proton and Electron are Made of Neutrinos

As a result of these impacts, the massive neutrino is rotated so that the velocity is perpendicular to the thrust, see below.

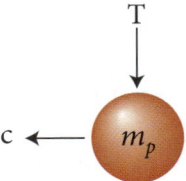

Neutrino Thrust Perpendicular to Velocity

The massive neutrino takes a circular path as shown here.

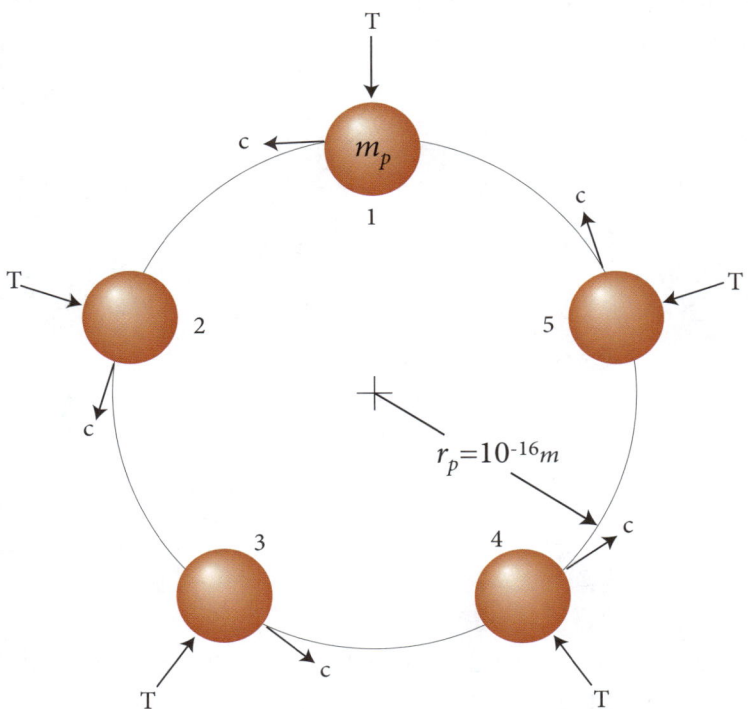

The Proton

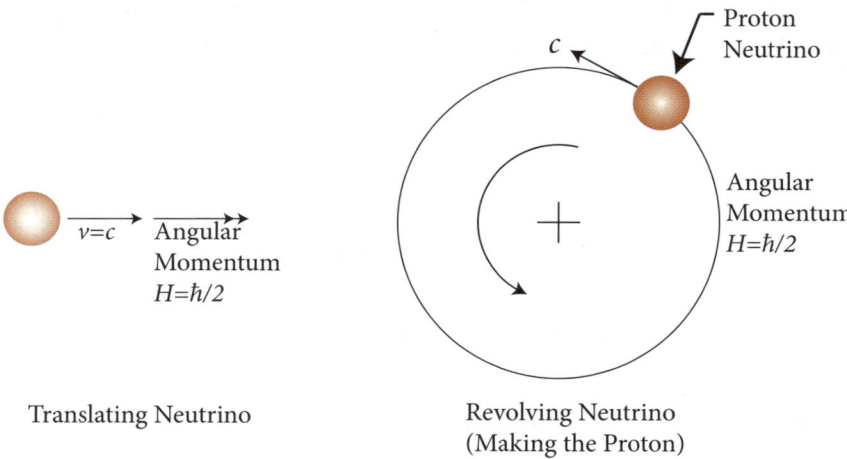

Translating Neutrino

Revolving Neutrino
(Making the Proton)

The thrust balancing the centrifugal force in the proton is shown below.

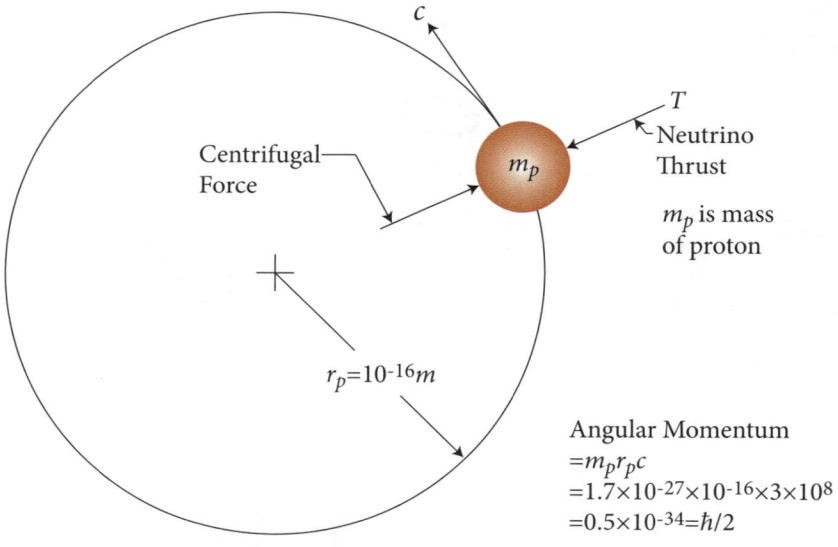

All neutrinos have the same thrust. There is only one value of mass which will balance the thrust and produce the same angular momentum which the neutrino had before getting knocked into its circular orbit.

The Proton and Electron are Made of Neutrinos

Thus, we know what causes the proton to have the mass it has. The proton is shown below.

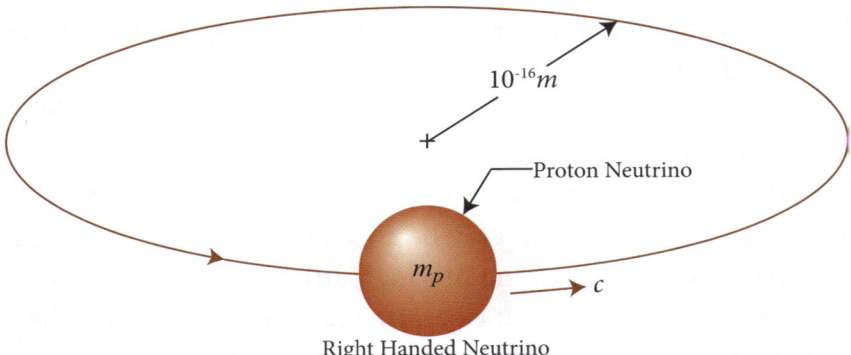

Angular momentum conservation is exemplified by a twirling skater, see the figure.

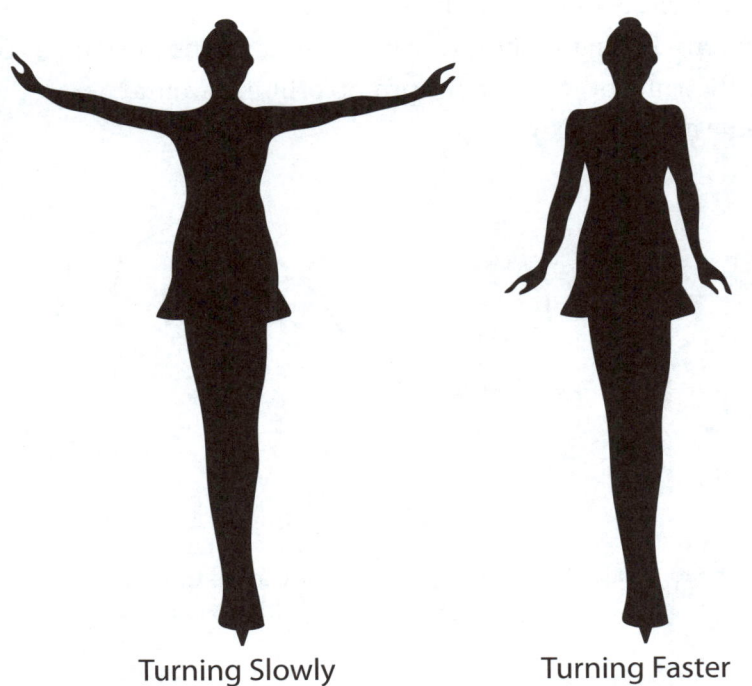

Angular momentum is *mass × distance* from *center of rotation × velocity*. The angular momentum for the left figure is that due to the body plus the arms. With the arms retracted, the distance from the mass center to the center of rotation reduces so that the angular velocity increases.

Each neutrino has the same value of angular momentum independent of the mass in the core. The angular momentum is due to the background gas coming into the neutrino and rotating around the axis of translation. The angular momentum of the core is negligible compared to the flowing particles in the compression chamber. This rotation for the neutrino making the proton is right-handed and is signified by a double headed arrow pointing in the same direction as the translatory vector.

When the proton neutrino is knocked into its circular orbit, the angular momentum which was parallel to the translational vector becomes a circulation around the path taken by the neutrino making the proton. The figure shows the proton-sized neutrino with its twist ω and the same neutrino in the proton orbiting around point A with the same amount of twist.

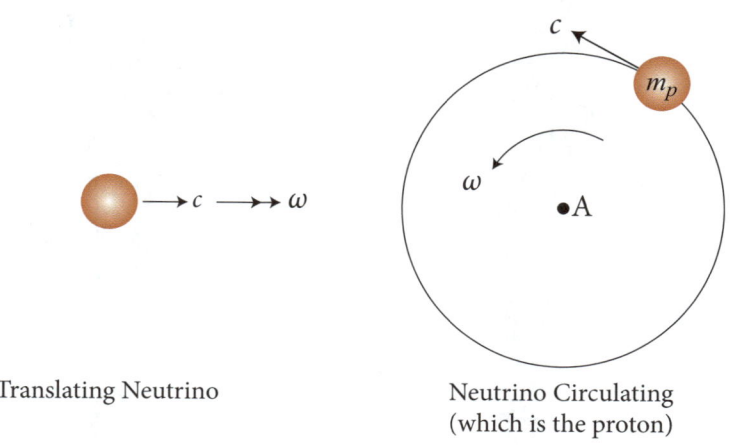

Translating Neutrino

Neutrino Circulating
(which is the proton)

The Proton and Electron are Made of Neutrinos

When the Proton Is Made a Circular Flow is Created

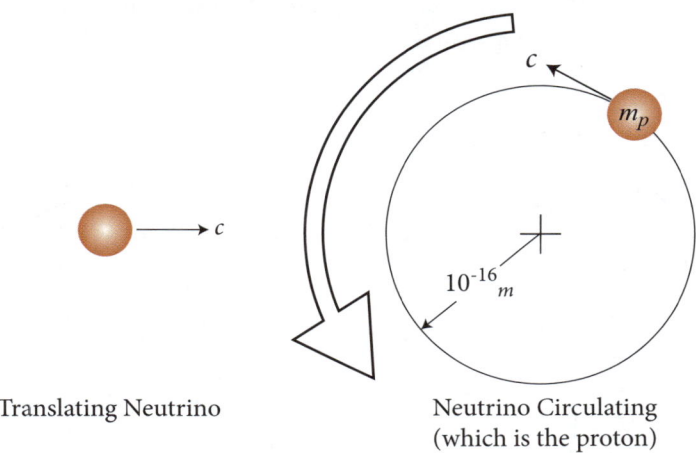

Translating Neutrino

Neutrino Circulating
(which is the proton)

When a neutrino circulates counterclockwise as in the above picture the background circulates in the opposite direction.

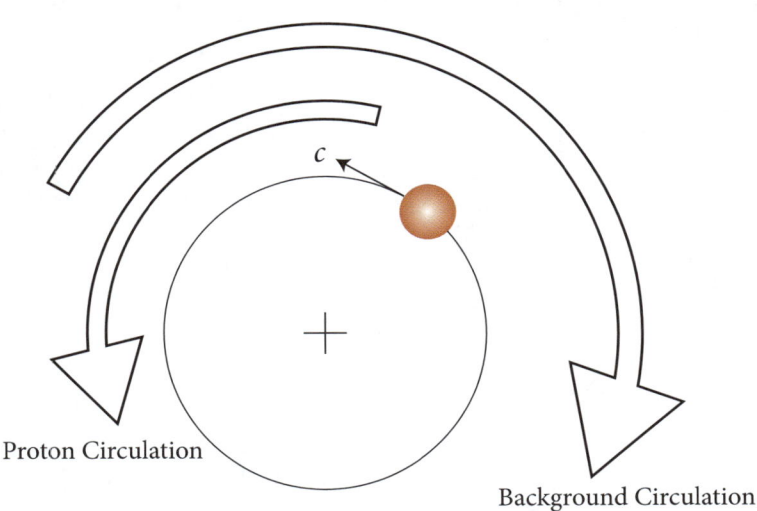

Proton Circulation

Background Circulation

The outer arrow shows the background circulating in the opposite direction. Angular momentum is always balanced.

When the proton is made, angular momentum produced must be counter balanced by opposite angular momentum. This process ends up making an electron.

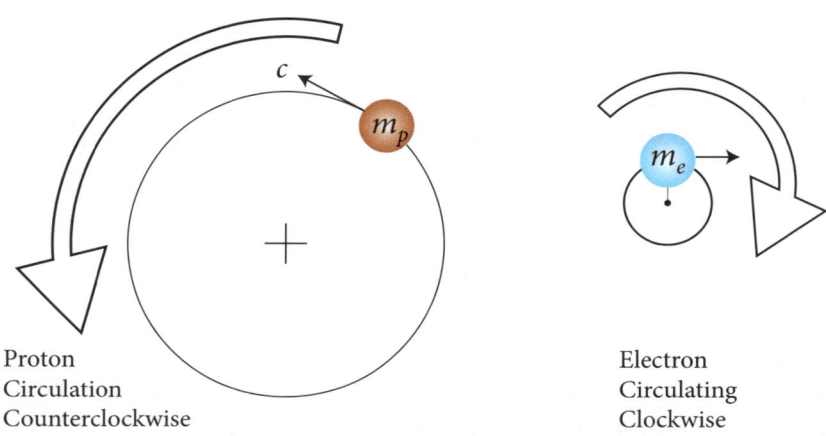

Proton
Circulation
Counterclockwise

Electron
Circulating
Clockwise

The proton makes a template for making an electron from the background gas.

The Proton and Electron are Made of Neutrinos

The electron is made of an electron-sized neutrino with left handed twist taking a circular path. The electron is shown below.

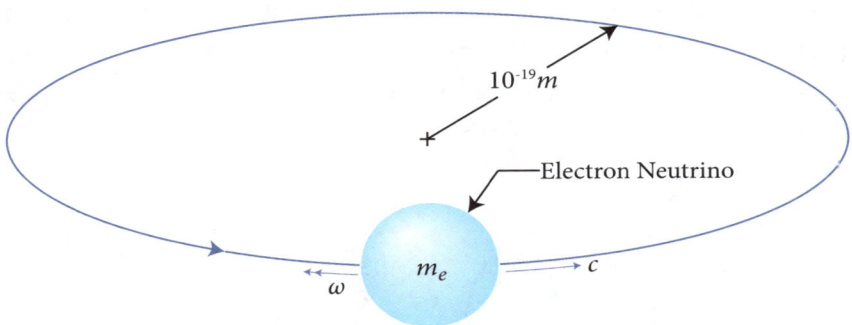

The electron neutrino is actually formed from the background (free) brutinos as a result of the template consisting of the proton formation mechanism.

The electron also must produce the electrostatic field like the proton's but of opposite *polarity*. The opposite polarity is provided by the electron neutrino being left-handed while the proton neutrino is right-handed. The two loops are shown below.

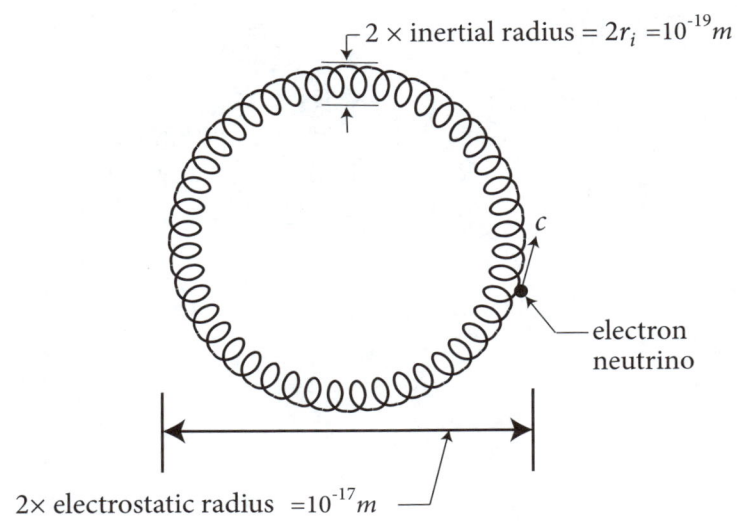

The electron electrostatic radius of $10^{-17}m$ is less than the proton radius of $10^{-16}m$ since the electrostatic pulse length produced by the slower velocity electron must take the same time duration as the proton-produced electrostatic pulse. Finally it is necessary that the electron have angular momentum of $\hbar/2=1.05\times10^{-34}/2=0.55\times10^{-34}$ $kg\text{-}m^2/s$. This requirement is met by the third loop – the angular momentum loop. See the figure.

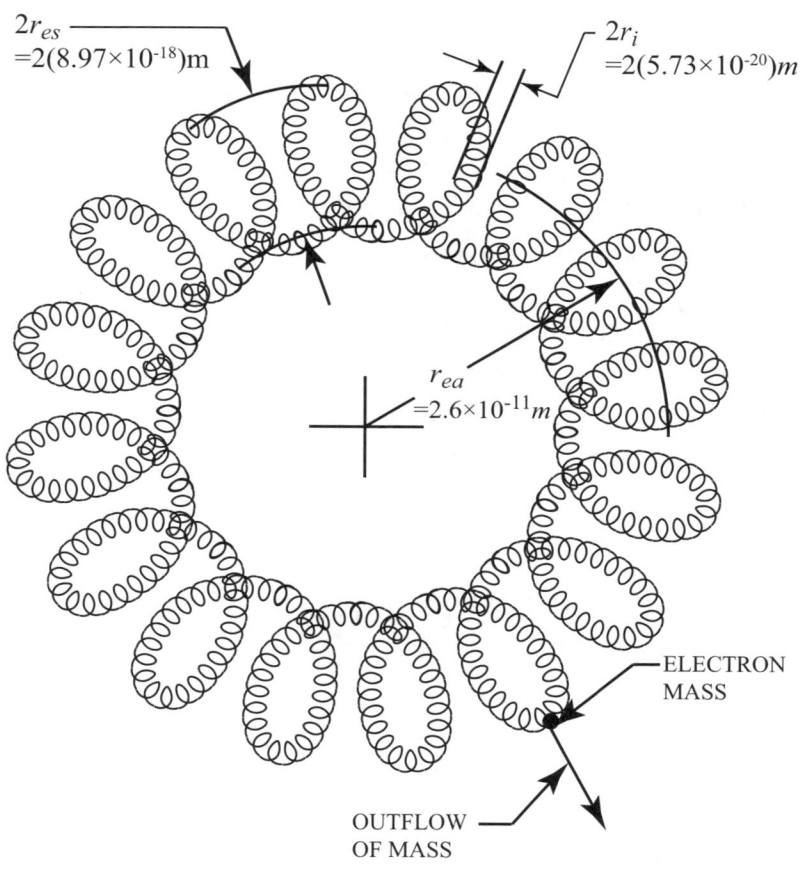

The Proton and Electron are Made of Neutrinos

The electron in the hydrogen atom thus may appear as below

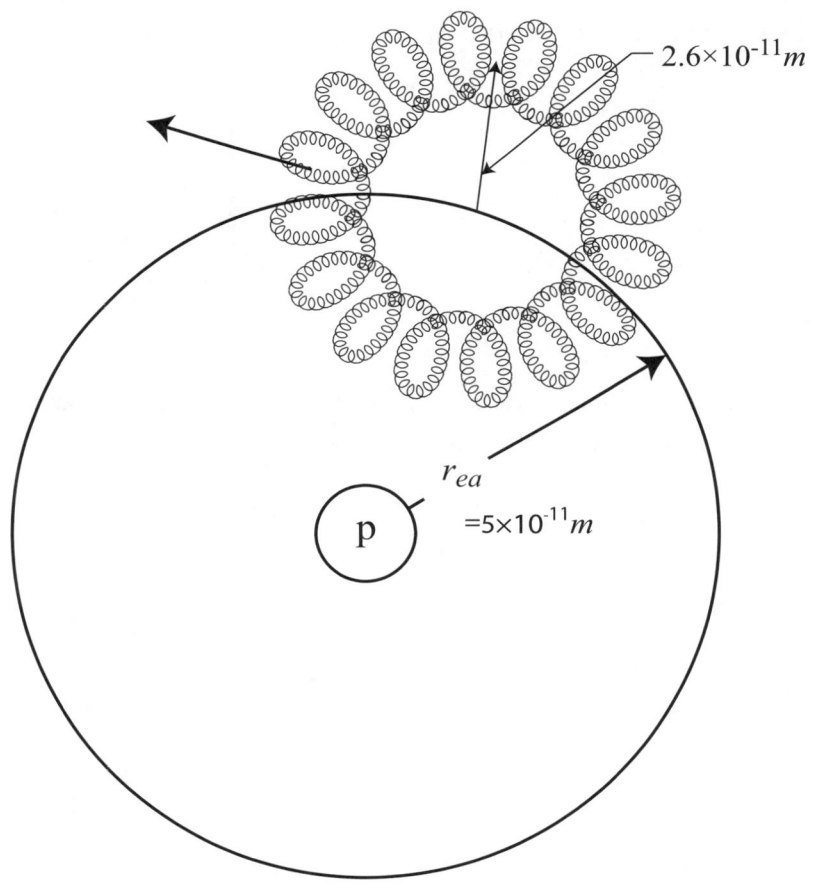

Another interesting aspect of electrons is their elusiveness. Niels Bohr assumed for hydrogen that the electron was a point mass orbiting the proton. Later theoretical work indicated that the electron was a *cloud* or something for which only a probability of locating the electron could be specified. Actually, the electron is a mass just like the proton except its angular momentum is opposite the proton's and its mass is 1/1836th that of the proton. However, while the resting proton neutrino takes a circular path, the resting electron takes a *triple looped* path. The smallest path is approximately a circle with a radius 1/1863th that of the proton orbital radius, see the figure.

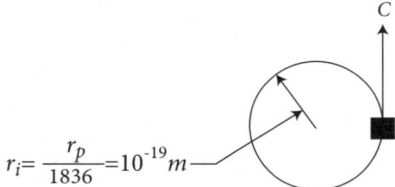

We call this the inertial loop and its radius is r_i. We also have the electrostatic loop and the angular momentum loop, see page 74.

5. The Electrostatic Force is the *Glue* Binding an Electron and Proton to Produce a Hydrogen Atom

When the proton neutrino travels around its circular path there is an in-out pulsation of the ether gas which when measured at some distance from the proton has essentially a spherically symmetric, radial in-out component. There is also a traveling wave in the direction tangential to a sphere whose center is at the proton center. The following figure shows the proton electrostatic field.

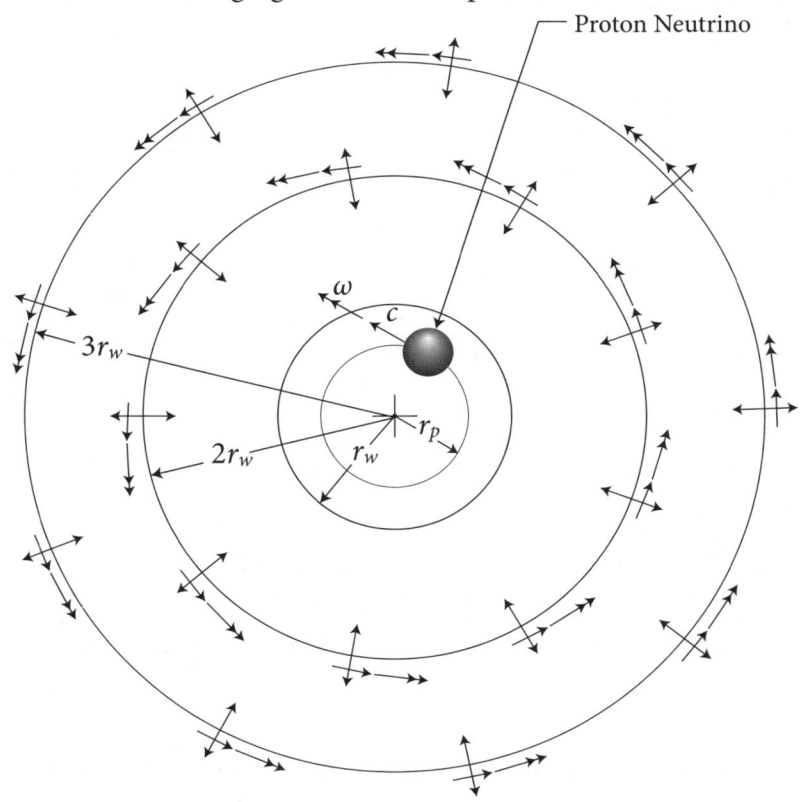

Proton Electrostatic Field
The Proton Makes a Radial Pulsing Flow
and a Transverse Traveling Wave
with Right-Hand Twist

The *field* of the electron is shown below.

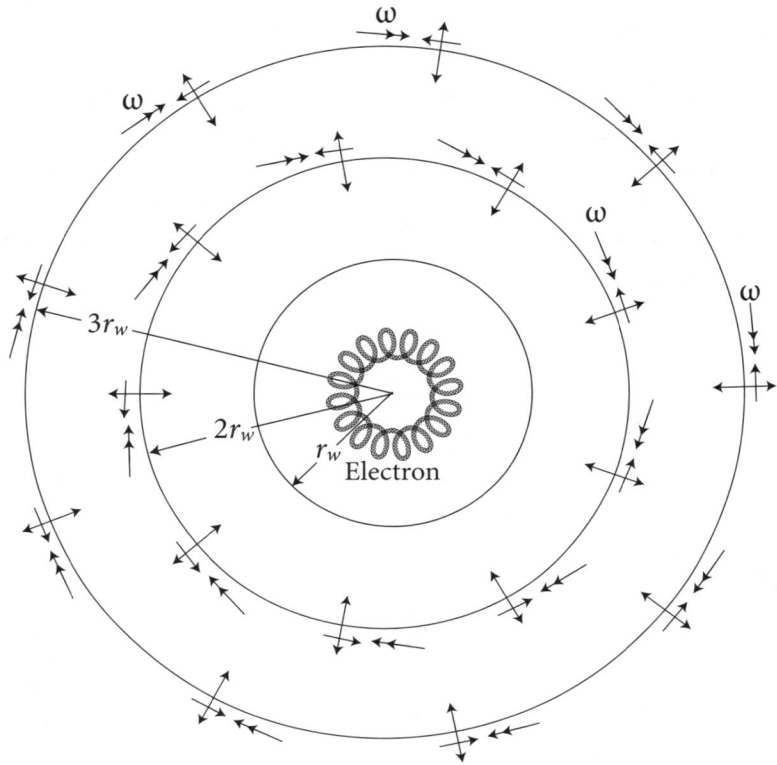

The Electron travels in a more complex path than the proton
Makes a radial pulsing flow
Makes a transverse traveling wave with left-hand twist.

The radial in-out vectors are shown above by vectors with arrows at both ends. The tangential motion is indicated by the single-headed vectors. The vector with the double arrows at one end represents rotation of the flow field. The rotary flow is induced by the left-hand rotation ω of the neutrino about its own axis.

The Electrostatic Force is the Glue Binding an Electron and Proton to Produce a Hydrogen Atom

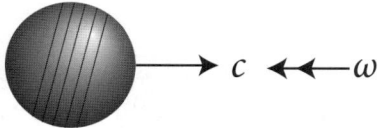

Left-Handed Neutrino

The electron is produced by a left-handed neutrino. Its geometric path is more complex than the proton but at a distance the flow oscillation is the same as that of the proton except for the difference in handedness, see the figure on the foregoing page.

It seems quite mysterious that a particle, such as a proton, buried in a gaseous *sea* could repel another proton at a large distance or could attract an electron which is far distant. This phenomenon, known as *action at a distance*, was experimentally proven almost a century and a half ago. Father and son researchers named C.J. and D.K. Bjerknes in Belgium (in 1877) demonstrated that *breathing spheres* immersed in water could attract as well as repel each other. Two identical elastic spheres were immersed in a large vat of water. The spheres could be symmetrically expanded and contracted much as a balloon can be expanded and contracted. Such an expansion-contraction motion is called the *breathing mode* for spheres. There were provisions for measuring the force tending to pull the spheres together or to separate them. When the spheres were breathing-in-phase (i.e., expanding at the same time and contracting at the same time) they attracted each other. When 180° out-of-phase they repelled each other. The electron and protons electrostatic fields act similarly except that the fields have *twist* components which provide for phasing.

There is another difference between breathing spheres immersed in water and electrostatic charges. Water acts like an imcompressible medium while the ether is a gas. For a gaseous medium breathing in-phase produces repulsion while out-of-phase produces attraction.

The effective breathing sphere analysis of interaction of two charged particles displaying the breathing sphere characteristics yields their interactive force. Like charges repel and unlike charges attract.

From the forgoing, two protons separated by a large (several proton diameters) distance r will repulse each other. The force of repulsion F_e is given by

$$F_e = \frac{e^2}{r^2}$$

where e is the fundamental electrostatic charge, its value is $e=1.4\times10^{-14} kg^{1/2}-m^{3/2}/s$. If the charges are produced by two opposite handed neutrino matter particles then the force will be attractive. The twist of the neutrinos produces the *sign* of the charge. Arbitrarily protons are assigned positive charge and electrons negative charge.

The simplest hydrogen atom consists of an electron orbiting a proton. The electrostatic force is attractive, $F_e=e^2/r_e^2$. The centrifugal force is $F_{cent}=m_e v_e^2/r_e$ which tends to make the electron take a straight path. In this expression m_e is the mass of the electron, and r_e is the orbital radius.

The Electrostatic Force is the Glue Binding an Electron and Proton to Produce a Hydrogen Atom

The figure shows the hydrogen atom.

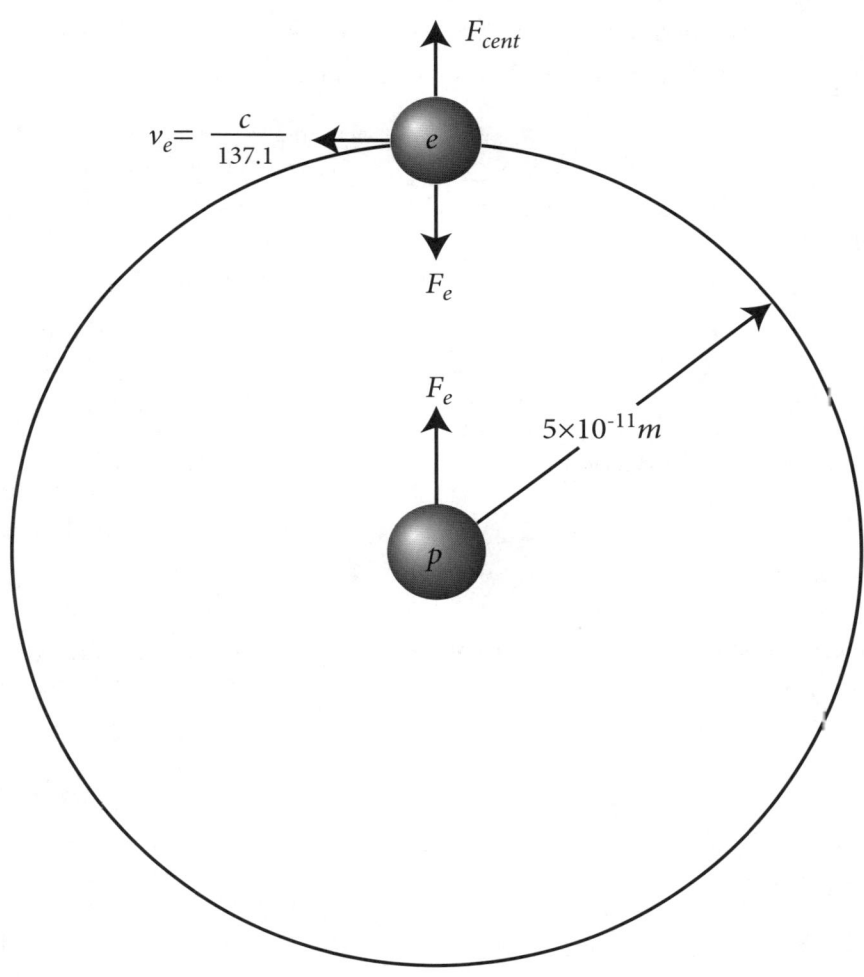

Hydrogen Atom

Previously we noted that an electron formation is much more complicated than the proton. When a proton is formed the flows produced by the proton act as a template to produce the electron using the background gas as its material.

The nutrinos in the hydrogen atom are shown here. We show a simplified representation of the electron path.

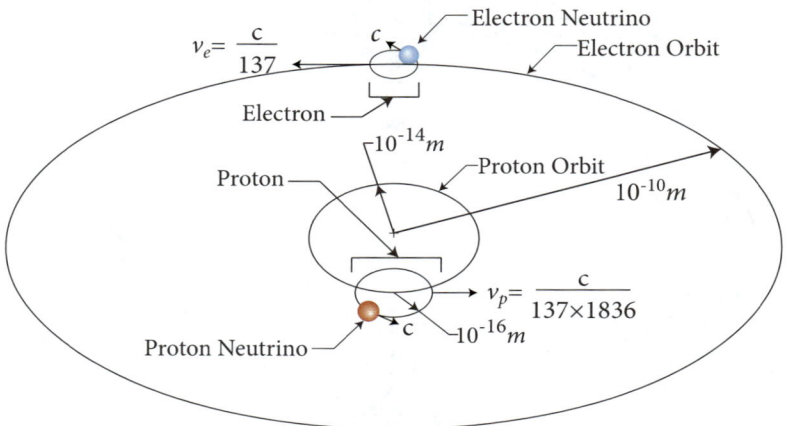

Here we show the neutrino orbits making up the electron and the proton.

The Electrostatic Force is the Glue Binding an Electron and Proton to Produce a Hydrogen Atom

The force fields producing the forces holding the electron to the proton are anything but uniform. Let us re-examine the flows produced by the neutrino making the proton. Let us look at the output from the proton neutrino, see the figure.

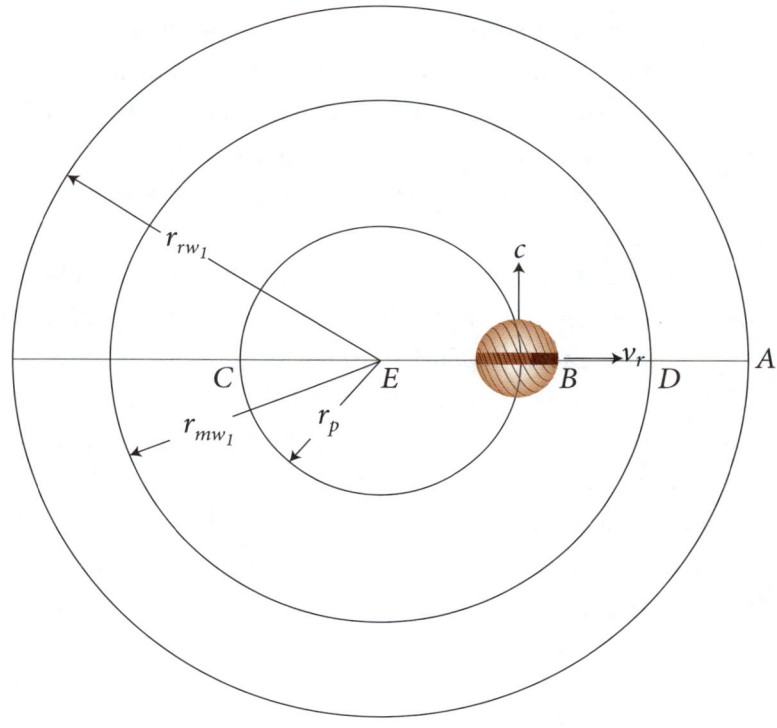

The Proton

The output from the front of the neutrino is at velocity v_r. An observer at D sees the outermost extent of the first wave whose radius is r_{rw_1}. What occurs is that when the proton neutrino is at B it sends out gas which at a later time will reach the maximum distance from B. As the neutrino moves toward C the flow begins to be radially inward.

When the neutrino reaches point C a gas flow directed toward D is generated and the outward flow at the slower velocity

83

v_m reaches a maximum at a shorter distance from the proton center and at some time later. The maximum is reached at a radius of r_{mw_1} — at a later time than when the maximum at A was reached. We have shown only the extent of the first wave due to the leading end of the neutrino and the first wave due to the trailing end. The neutrino generates these spherical waves which extend very far from the neutrino – the maximum velocities continue at the values v_r and v_m but the mass flow rates diminish as the square of the distance from the proton. Eventually the flow rates get so small that just one brutino will flow and that is the extent of the electrostatic field – the distance may be light years from the proton.

When a proton is first isolated from the orbiting electron in the hydrogen atom, i.e., when the hydrogen atom is *ionized* the field component due to v_r expands radially at the velocity v_r and the component due to v_m similarly expands radially at velocity v_m. Since the fields are expanding at different velocities there are distances from C where they are exactly in-phase or exactly out of phase. We show a plot of velocity versus distance r from B taken at a given time and at a location such that the two waves are exactly in-phase at the left side of the diagram.

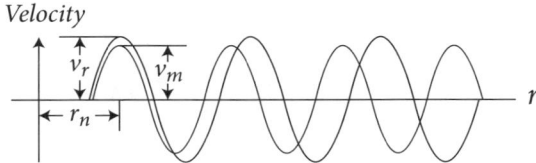

At distance r_n the two waves are exactly *in-phase* — they are *resonant*.

At some greater distance they can get nearly completely *out-of-phase* (as shown at the right of the figure) and doubling that distance they can get nearly in-phase, i.e., the next resonance location. The number of cycles from one resonance to the other is $v_m/(v_r - v_m)$. This can be seen from the following figure showing

The Electrostatic Force is the Glue Binding an Electron and Proton to Produce a Hydrogen Atom

peak upward velocities.

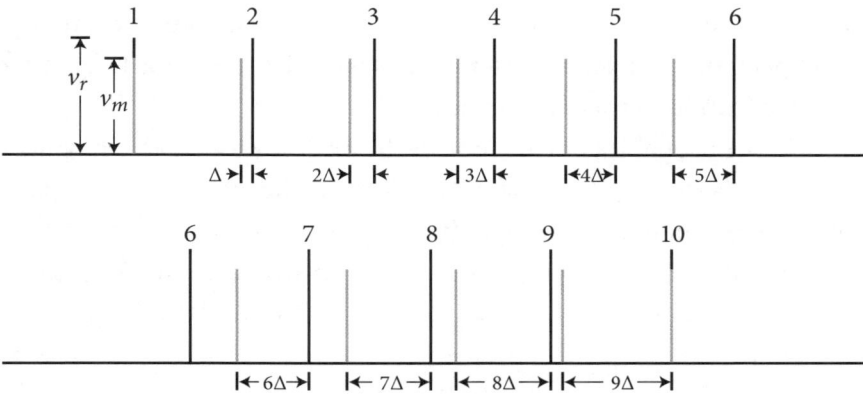

Here we let Δ be the amount the v_m curve lags the v_r curve for each sycle. Letting τ be the time for one cycle of the proton neutrino we have

$$\Delta = (v_r - v_m)\tau$$

For this example we let v_r be $1.1 v_m$ thus in 10 cycles the two waves are in-phase again. We now determine that the frequency of resonance is

$$f = \frac{v_r - v_m}{v_m} = \frac{1.1 v_m - v_m}{v_m} = 0.1$$

or, in general

$$f = \frac{v_r - v_m}{v_m}$$

Knowing that the electron must produce an electrostatic field which balances the proton field, in so far as possible, we assume that the electron electrostatic field will have the same resonances.

Let us define a *wavespace* as the size of each wave produced by the proton (or electron) neutrino. We recognize the radial extent of this space as the distance from one wave to the next. Similarly, we expect that tangential oscillations would produce a tangential distance from one wave to the next.

The probability of any one of the proton wave spaces having a resonance is $(v_r - v_m)/v_m$. The probability for the electron wave space having a resonance likewise is $(v_r - v_m)/v_m$. The probability of an electron and proton having a resonance in the same wave space is $[(v_r - v_m)/v_m]^2$. Since $v_r - v_m = \sqrt{3\pi/8}$ we note that this value is

$$[(v_r - v_m)/v_m]^2 = 1/137$$

We note that this value is close to the value of the fine structure constant.

Now if the electron electrostatic waves could pass through the proton electrostatic waves unimpeded they would move at the speed of light. Possibly what is happening is the field relative motion only occurs in those wavespaces where there is resonance. Thus the electron velocity would be

$$v = [(v_r - v_m)/v_m]^2 c = c/137$$

And this would be the square of the coupling velocity (or the coupling constant) for quantum electrodynamics.

Clearly the above arguments are quite conjectural. Much remains to be learned of the interaction of electrostatic fields. However, we do know that each electrostatic field is *wavy* with resonances at approximately every tenth wave. It is yet to be determined how these resonances control electrostatic field interactions.

6. The Proton Electrostatic Field Mixed with the Electron Electrostatic Field Produces Gravity

We cannot see brutinos, electrons, protons, hydrogen atoms, and larger atoms. What we can see is groups of trillions of trillions of atoms. A trillion trillion atoms of iron makes up about one kilogram of iron. We should be able to see a grain of iron with a mass of a billion trillion atoms. How are things we see made? We have seen how neutrinos and then hydrogen atoms are made. The first thing made that we could see is a *small* group of several billlion trillion hydrogen atoms. Hydrogen atoms are made continually and they begin to assemble because each hydrogen atom has a gravitational *field*. What is the mechanism which produces gravity?

Recall that the electrostatic field of the proton is the perturbation in the ether background gas produced by the proton neutrino moving around its circular path with a path radius of r_p ($=10^{-16}m$). See the figure.

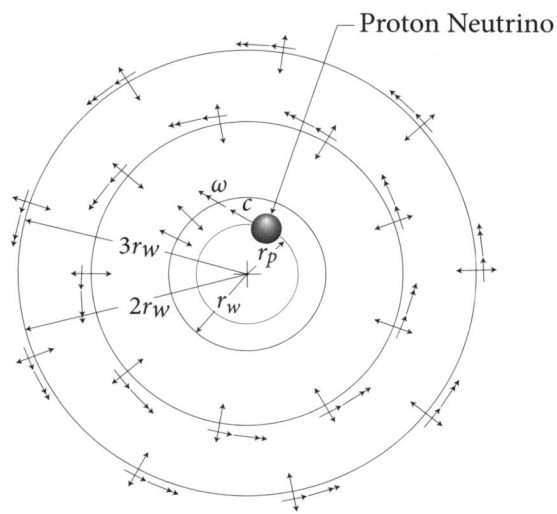

Proton Electrostatic Field

The Proton Makes a Radial Pulsing Flow and Makes a Transverse Traveling Wave

The electron produces a similar field with the same radial pulsing flow but since the electron neutrino is left-handed the tangential traveling wave is left-handed as shown in the following figure.

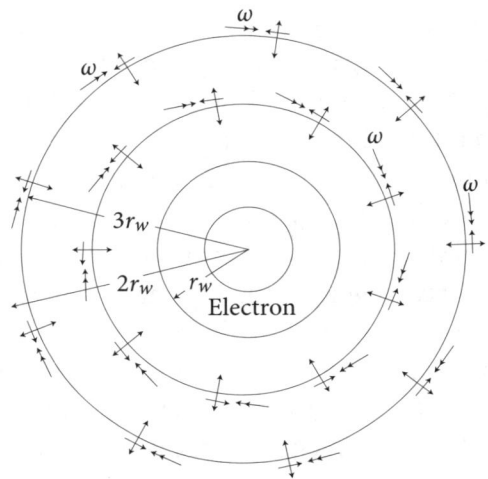

Electron Electrostatic Field

The Electron Makes a Radial Pulsing Flow and Makes a Transverse Traveling Wave

The Proton Electrostatic Field Mixed with the Electron Electrostatic Field Produces Gravity

Let us now look at the distant fields of the proton and electron side-by-side.

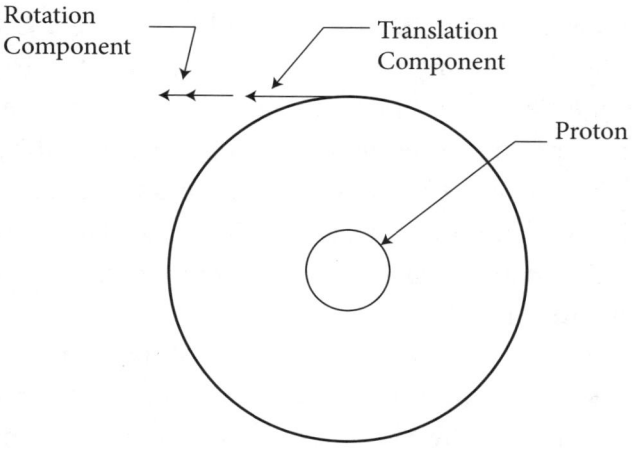

Proton Distant Field

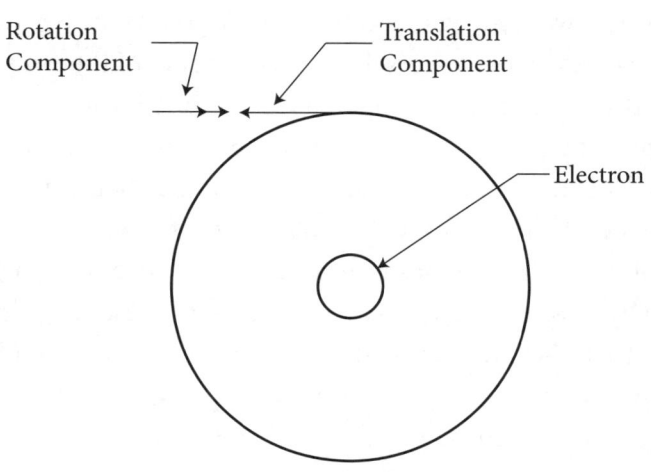

Electron Distant Field

In the hydrogen atom the two flow fields are superimposed on each other. The opposite rotary components cancel each other. In order for the translatory components to be nulled the complete assemblies rotate and translate with respect to each other – as exemplified by the orbit of the electron about the proton. By doing this the distant flow field of the electron is almost exactly nulled out by the distant flow field of the proton. They are *out-of-balance* just by the amount required by the size of the brutino. The end result is that the distant field is a radial oscillation with a half amplitude equal to the radius of the brutino. This radial oscillation is the essence of the gravitational field of the hydrogen atom.

As you can see from the foregoing discussion there is much to be learned about gravitation. However, it is certain that gravitation is produced by the superimposition of the positive and negative electrostatic fields produced by the proton-electron pair in the hydrogen atom, or in the neutron.

The electrostatic field of a charged particle spreads out over the complete surface of any sphere with its center at the charge center. Thus, the strength of the field decreases inversely with the square of the distance from the particle. Now, the gravitational force between two hydrogen atoms, just like the force between two charged particles, varies inversely with the square of their separation distance. If there are N_1 hydrogen atoms at one location and N_2 at another then the gravitational force will be $N_1 N_2$ times as much as for one hydrogen atom at each of two locations. Neutrons produce a similar electrostatic field as hydrogen atoms. Now, the mass of an assembly of hydrogen atoms and neutrons is proportional to the number of hydrogen atoms plus neutrons. From these considerations it is clear that the gravitational force between two assemblies of matter is proportional to the product of their masses and inversely proportioned to the square of their separation distance. Thus, the force of gravity F_g is

The Proton Electrostatic Field Mixed with the Electron Electrostatic Field Produces Gravity

$$F_g = G \frac{m_1 m_2}{r^2}$$

Where G is the proportionality constant – which is the Newtonian universal constant of gravity. The measured value of G is $6.673 \times 10^{-11} m^3/(kg^2\text{-}s^2)$.

The figure shows two hydrogen atoms attracting each other.

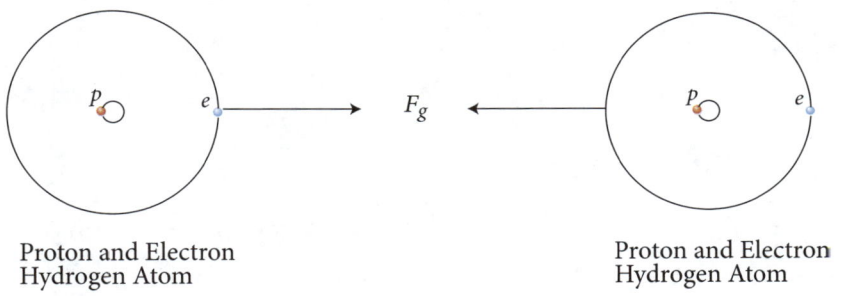

Proton and Electron Hydrogen Atom

Proton and Electron Hydrogen Atom

The Gravitational Force

7. Hydrogen Stars

Let us now explore the effect of gravitation. Hydrogen atoms are continually made throughout the universe. The figure shows brutinos, neutrinos, and hydrogen atoms, but not to scale.

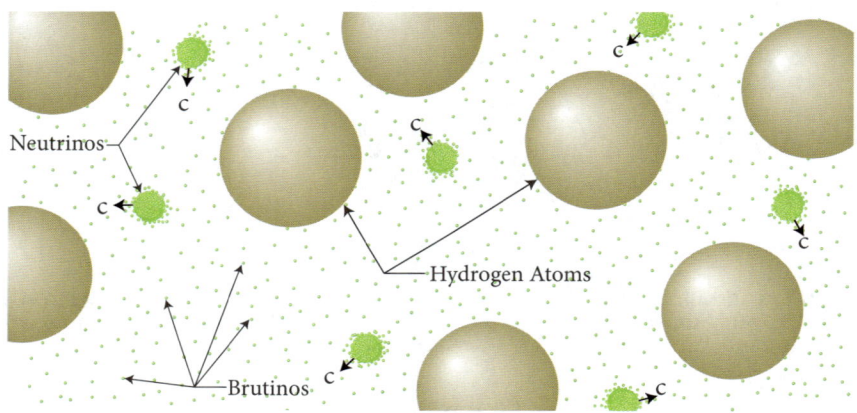

As a result of the gravitational force between hydrogen atoms, the atoms begin to form clusters, see the figure.

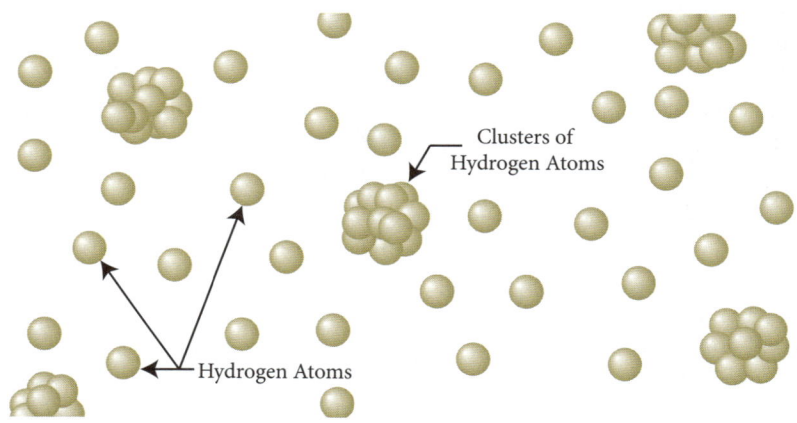

Hydrogen Stars

Large clusters of hydrogen atoms are formed.

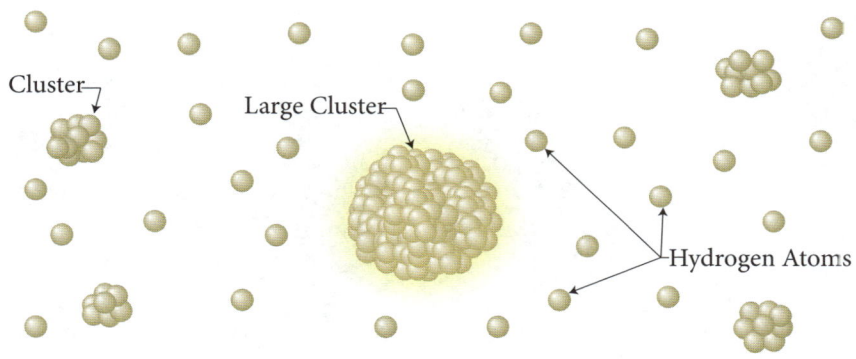

The clusters continue to grow until a small hydrogen star is formed, see the figure.

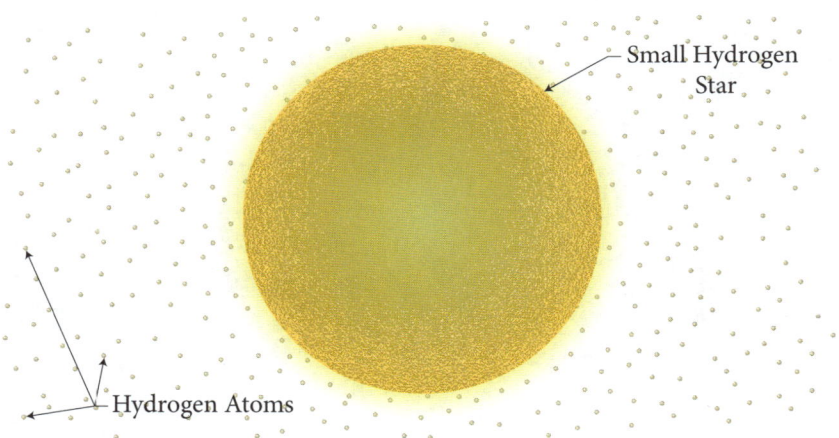

Large hydrogen stars are formed

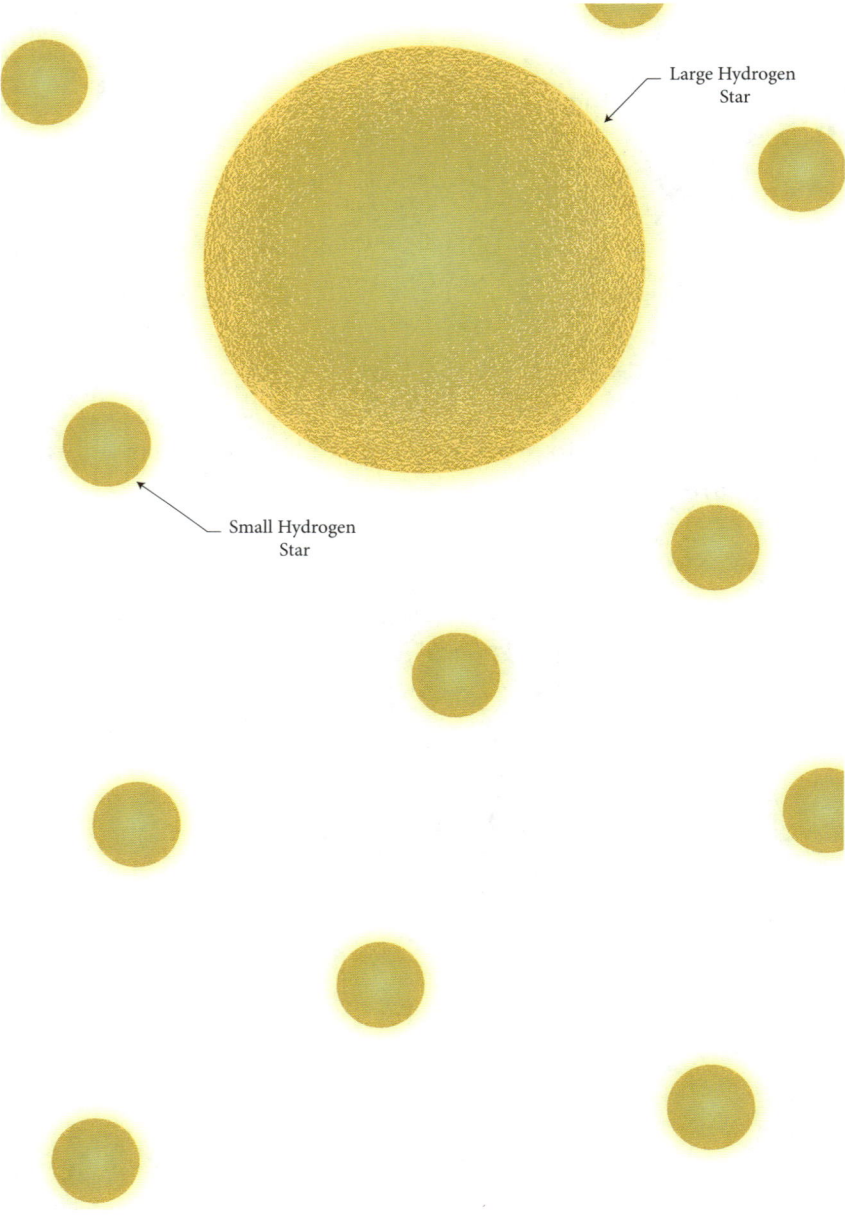

8. The Neutron and the More Massive Atoms

When hydrogen is subjected to large pressures the electron will be forced close to the proton i.e., to 1/100,000th the distance separating them in a hydrogen atom. The hydrogen atom then becomes a neutron. The electron must speed up to develop a large repulsive centrifugal force to balance the greatly enlarged electrostatic force. The figure shows the neutron. We note that the neutron consists of two orbiting neutrinos instead of one.

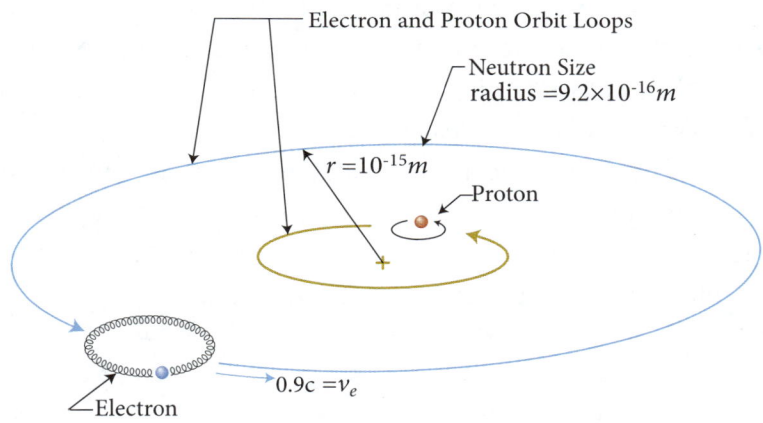

Large stars have large gravitational fields. These fields interact with the small gravitational fields of an individual hydrogen atom to attract it to the star. These individual hydrogen atoms are continually produced in space from the random motion of neutrinos. When the hydrogen atom gets *captured* by the star's gravitational field it is accelerated toward the star and can reach a very high energy level by the time it reaches the star. Some impacts will convert the hydrogen atom to a neutron. Some extremely large stars have gravitational fields which are strong enough to convert hydrogen at rest into a neutron. Neutrons are contiually made by large stars.

A group of hydrogen atoms and neutrons, under sufficient pressure, can produce an interaction between a free proton and the *proton* in the neutron. This interaction is the strong nuclear force – or the force involved in nuclear bombs.

The nuclear force is the result of the flows produced by the two individual neutrinos of the protons. For two neutrinos to interact to produce a force they must have the same mass so that their orbital diameters are the same, see the figure.

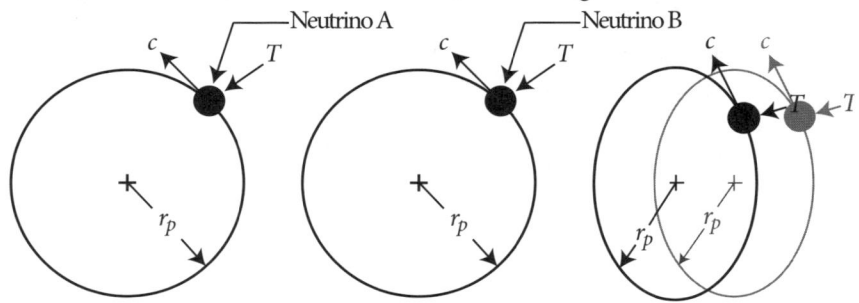

Two Interacting Orbiting Neutrinos

The flows that produce the force is indicated by the following figure.

The Neutron and the More Massive Atoms

As a result of the flow between the neutrinos, the static pressure p_o between the neutrinos decreases to a reduced static pressure p_r and the background pressure pushes the two neutrinos together.

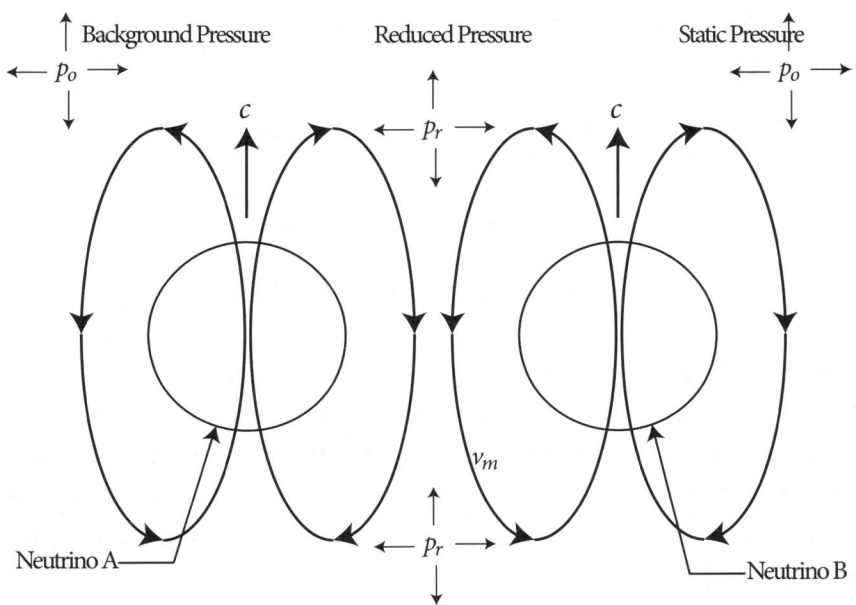

Interacting Neutrino Flows

    This pressure reduction as a result of flow is familiar to most people by the phenomenon of the cooling effect produced by the high speed flow of air from a nozzle connected to an air compressor. The cool air is produced by the reduced random velocities of the air atoms, which velocity causes the temperature, and this reduced temperature reduces the static pressure. The force between two neutrinos is on the order of 200 to 300 newtons. Since there are 10^{25} proton pairs in a kilogram of matter, there is a high total combined force holding them together. A nuclear bomb only utilizes a very small percentage of this force.

Two proton pairs also can interact with each other as shown by the following figure.

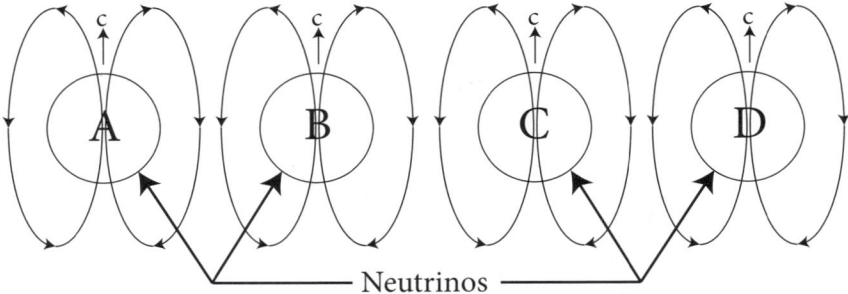

Two Pairs of Interacting Protons

Notice that the flows between A and B as well as C and D are downward so A is attracted to B and C is attracted to D. The flows produced by AB on its right and by CD on its left are both downward. Thus, pair AB will be attracted to pair CD. The result of the coupling of two pairs of nucleons (actually two pairs of proton neutrinos) produces helium.

More and more pairs can be assembled to produce larger and larger atoms. However, there is a limit to how many pairs can be assembled and remain stable. More distant nucleons produce flows at a given nucleon which can be opposite the flow produced by its immediate neighbor. However, if there are many nucleons surrounding the given nucleons then flows can add up and negate the flow gluing the nucleon to its immediate neighbor. The actual limit of nucleon pairs which can be bonded stably is over 100.

Incidentally when nucleons bond, some of the nucleons are neutrons and some are protons. Thus, as a larger atom is made the number of electrons orbiting at radii in the vicinity of 10^{-10} meters will be the same as the number of nucleons which are protons. Thus, in all atoms the number of protons will be equal to the number of electrons.

The Neutron and the More Massive Atoms

Atoms with numerous neutrons and other proton-electron pairs will attract each other by the gravitational force. The amount of attraction is measured by the number of proton-electron pairs. The force for an atom of helium interacting with another atom of helium is shown below. F_g is the gravitational force between two hydrogen atoms. The gravitational force between two helium atoms is $16\, F_g$.

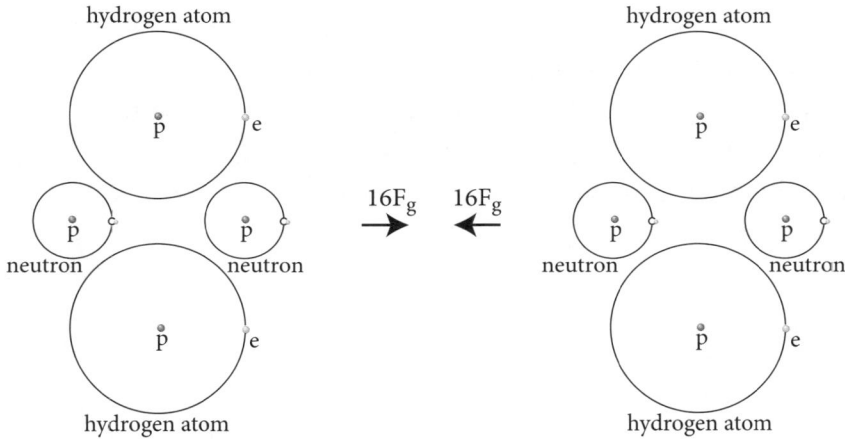

The Gravitational Force for Two Helium Atoms

The force of gravitation between two helium atoms is 4×4=16 times the force between two hydrogen atoms and is the same as between two assemblies of 4 hydrogen atoms each. However, even though the mass of a helium atom is slightly larger than the mass of 4 hydrogen atoms, we still use the equation

$$F=G\frac{m_1 m_2}{r^2}$$

for the gravitational attraction. What occurs is that G is measured for various assemblies of protons and neutrons and is not known with enough accuracy to distinguish between two all hydrogen assemblies and two all neutron assemblies — if such an experiment were possible

9. Neutron Stars and the Big-Big Bang Theory

Hydrogen stars get large enough so that their gravitational fields are sufficiently strong to develop large enough energies to transform hydrogen atoms to neutrons, see the figure

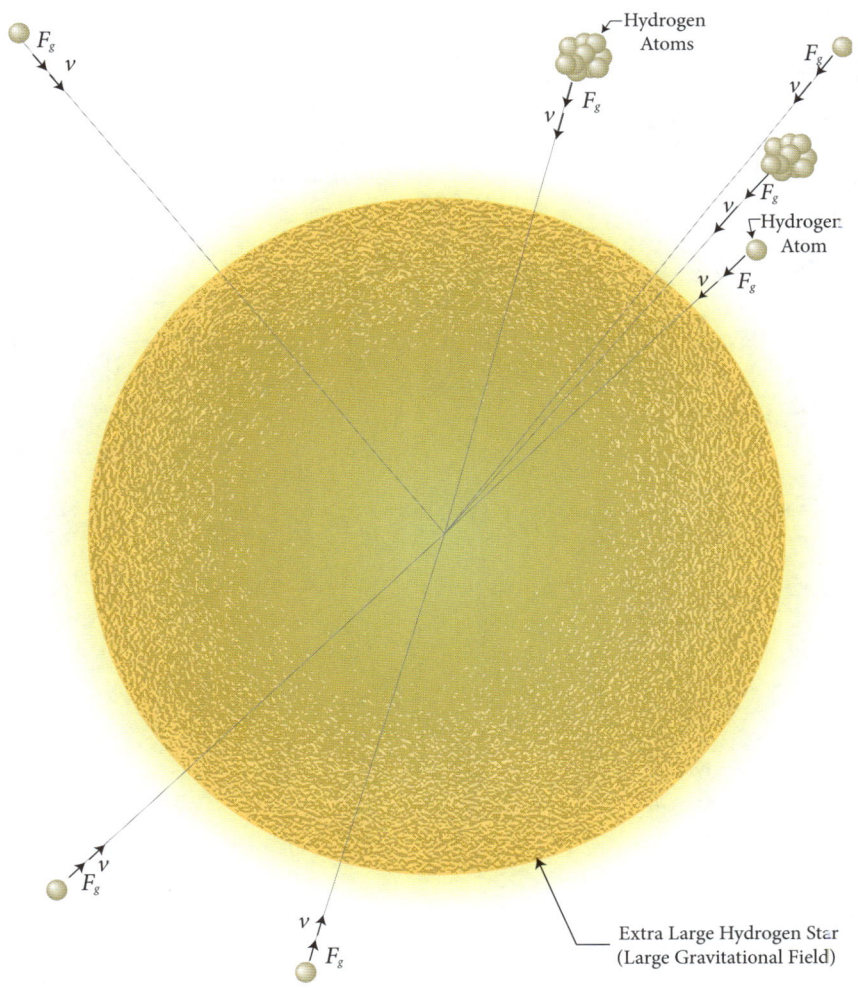

The electron orbit collapse is illustrated below.

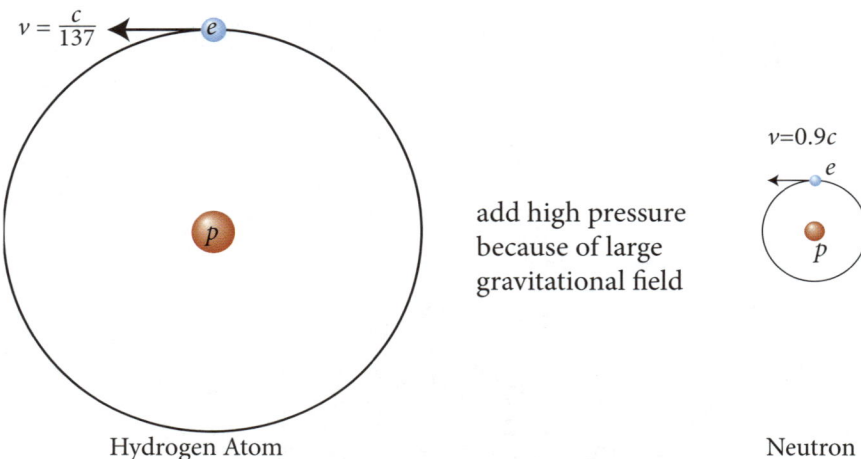

Heavier Atom Stars

Larger hydrogen stars with larger gravitational fields produce larger atoms and thus the hydrogen stars grow into heavier atom stars.

Heavier Atom Star

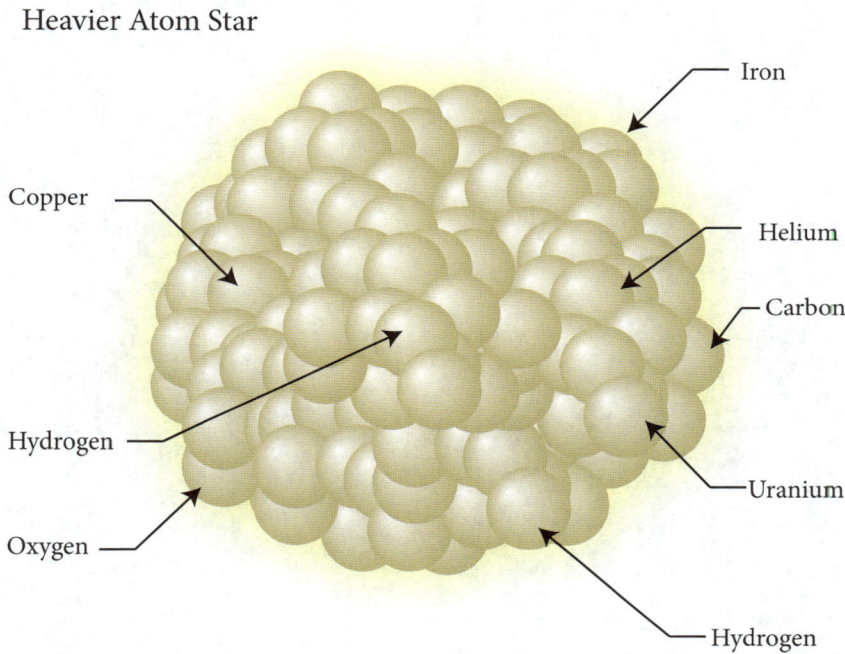

With the continual growth hydrogen stars become stars of all kinds of atoms. With continual growth the gravitational forces inside the star collapses the electronic structure of atoms and transforms them into neutrons. A neutron star then is born, see the figure.

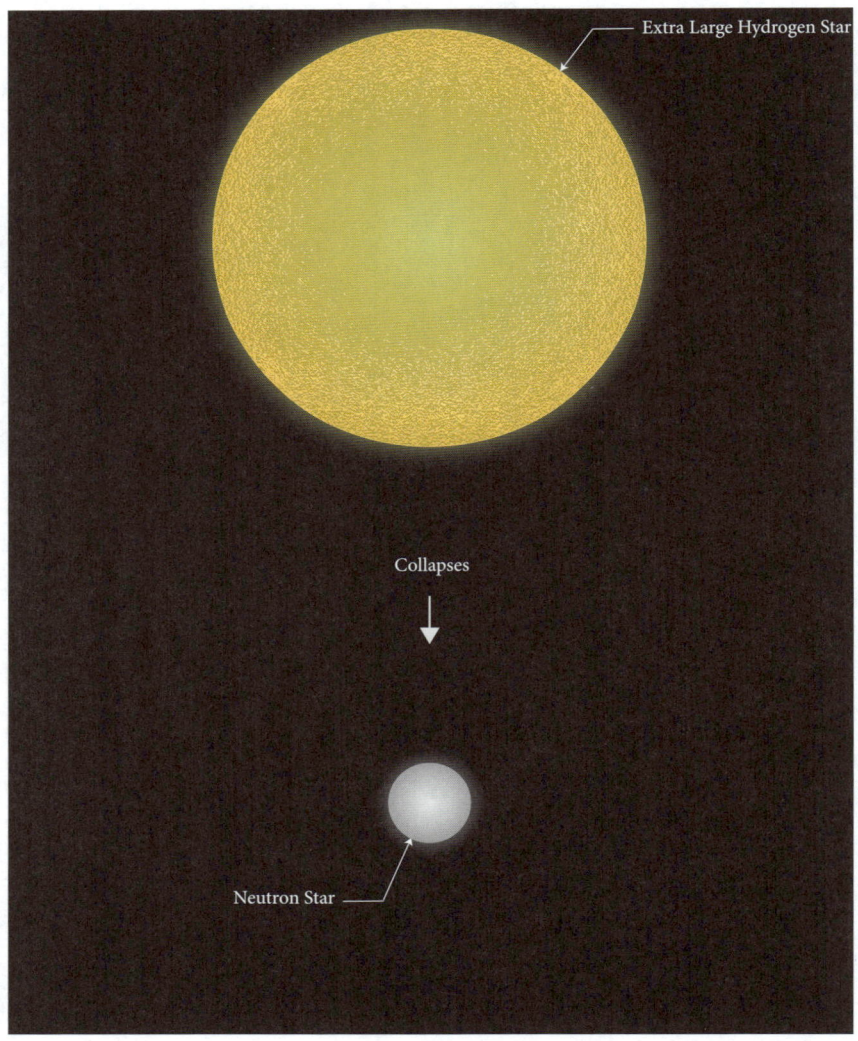

Neutron Stars and the Big-Big Bang Theory

The neutron star continues collecting mass from its surroundings.

Continued Growth of Neutron Star

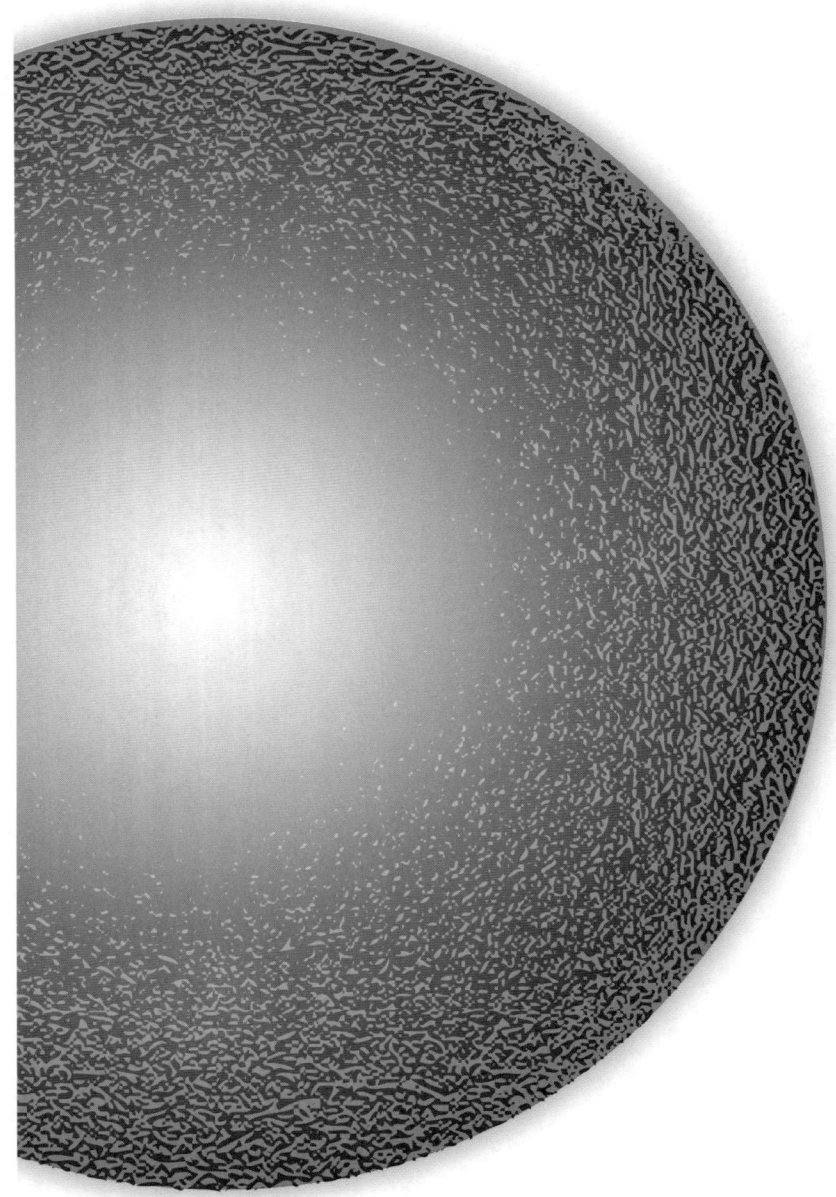

Extra Large Neutron Star

Neutron Stars and the Big-Big Bang Theory

The Big – Big Bang

Eventually a neutron star gets large enough (light years in diameter) so that its gravitational field collapses the nuclear orbits. As a result the neutrinos making protons and electrons take straight paths instead of circular paths.

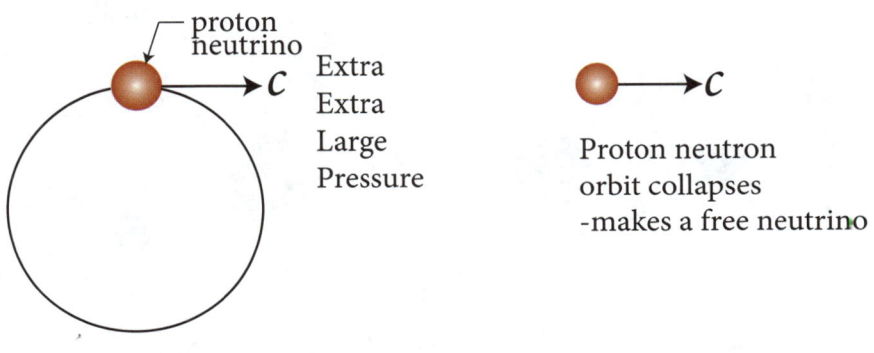

Many Proton Orbits Collapse

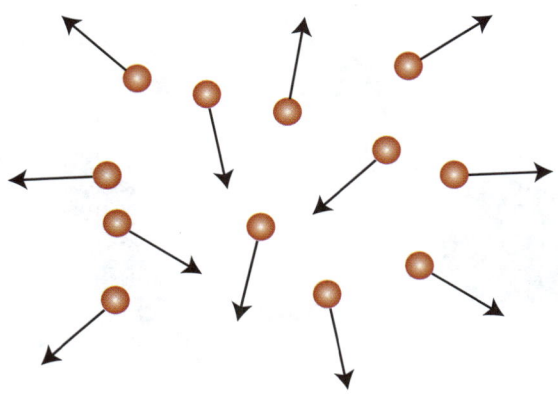

Neutrinos Go In All Directions
Causes Star to Explode

Physics for the Millions

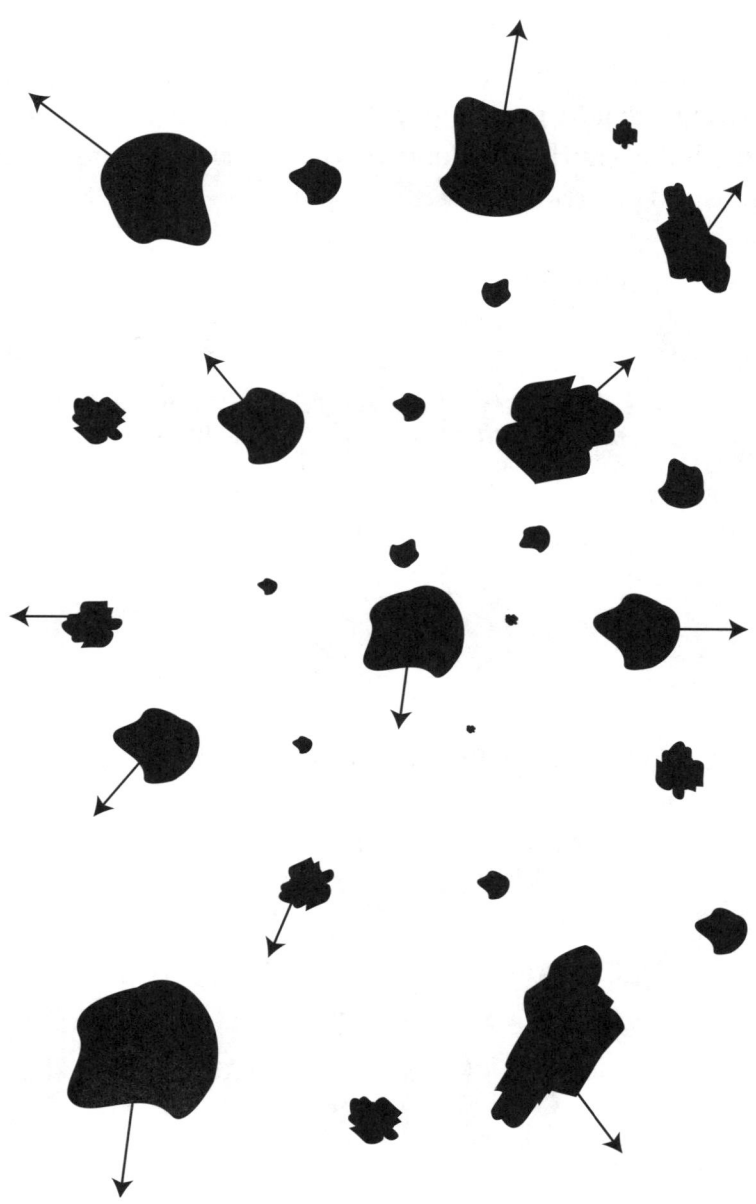

Exploding Neutron Star
Chunks of Matter Fly In All Directions

Neutron Stars and the Big-Big Bang Theory

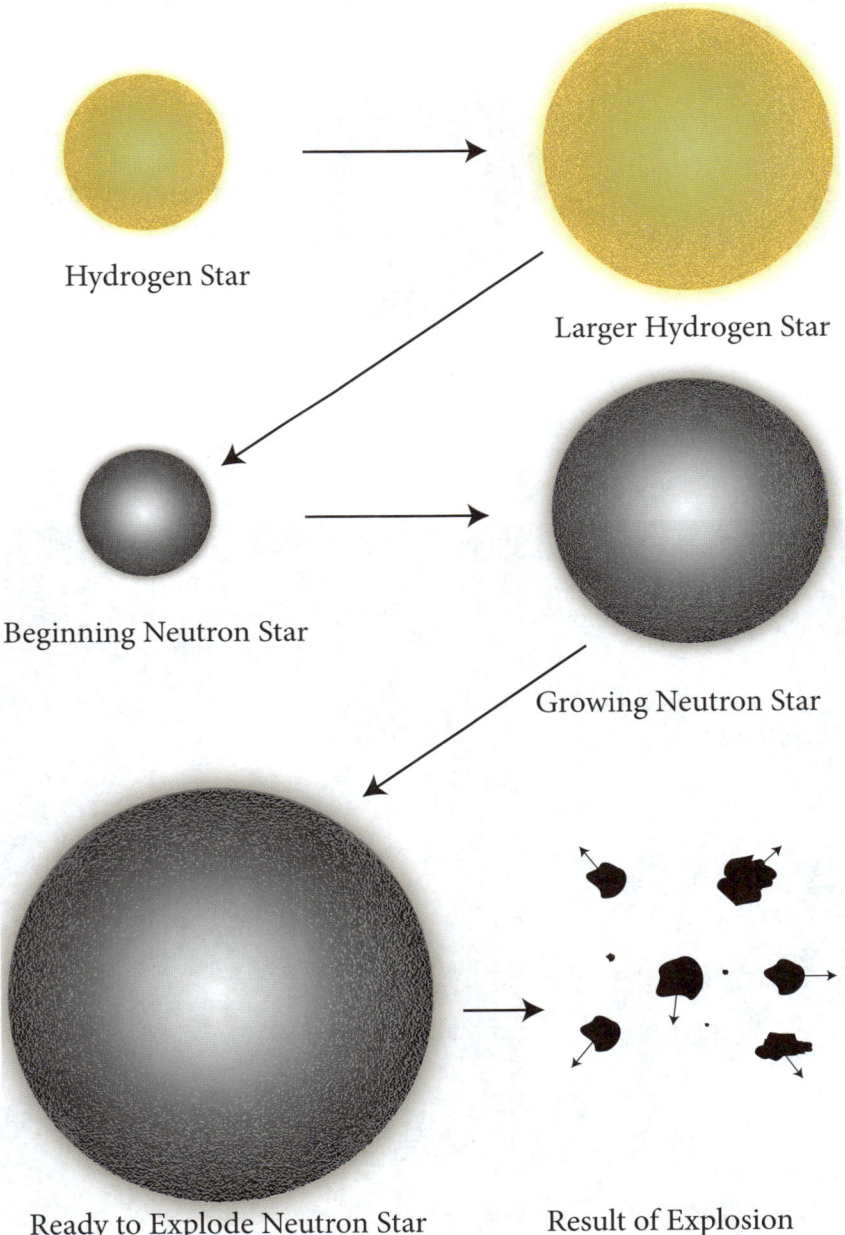

The neutron star explodes to make smaller stars, planets, moons, asteroids, and other space junk.

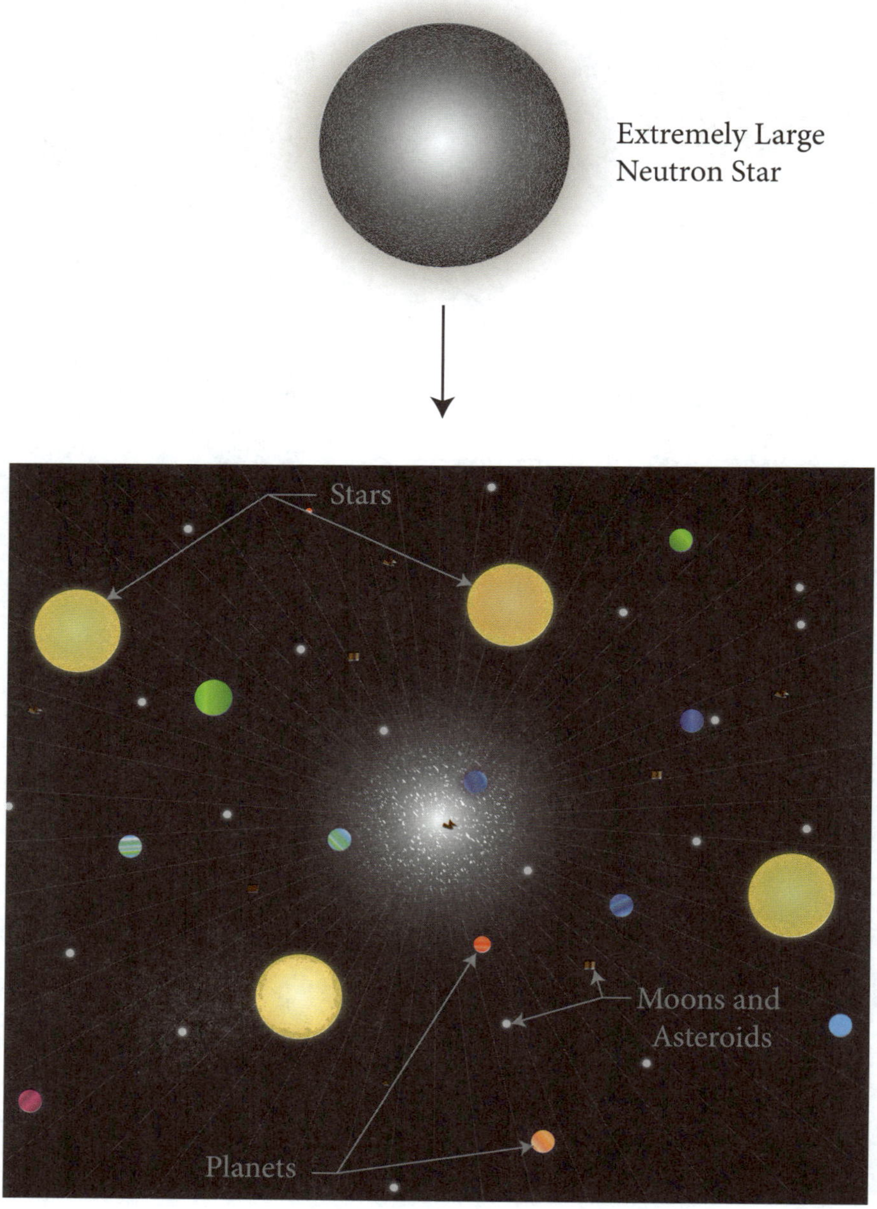

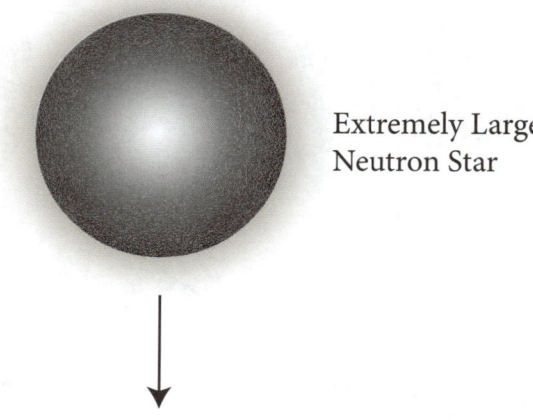

Extremely Large Neutron Star

Neutron Stars and the Big-Big Bang Theory

Just after the explosion the matter is all very hot, obviously, as a result of the explosion. It takes many years for matter to cool enough for living matter to originate. Here we show a star, a planet, a meteorite, and a comet.

One of the Neutron Star Explosion Fragments is the Earth

The large Neutron Star Explodes and Produces Fragments (Smaller stars, planets, moons, etc.) One of these fragments is the earth.

New Earth
Hot!

Time Passes

Cool
Earth

Time Passes

Bacteria
plants
and animals
appear

10. Photons Move Almost Everything We See

The previous chapters have shown what matter is and where it comes from. In this chapter we will show how we detect matter and how it changes. Detection and practically every change in the universe is facilitated by photons. Photons impinge upon our eyes and we see. The way we see is mostly a result of photons scattering (i.e., being reflected) from objects and then intercepted by our eyes. However, some objects, such as red hot pokers, emit photons which then intercept our eyes and we say the pokers are glowing and hot. Photons make up the light we see. The light we see looks like a continuum much as the pressure of a gas acting upon your body feels like the result of something continuous acting on your skin. But we know the pressure of a gas is due to repeated collisions of gas molecules with the skin. Light may look continuous but just like a gas, it consists of many, many particles.

Most of the photons on the earth come directly from the sun. The sun emits 10^{46} photons per second. The photons spread out from the sun in a spherically symmetric pattern. Thus since we are 10^{11} meters from the sun there are $10^{46}/(4\pi 10^{22})=10^{23}$ photons per second impinging upon each projected square meter of area of the earth, see the figure.

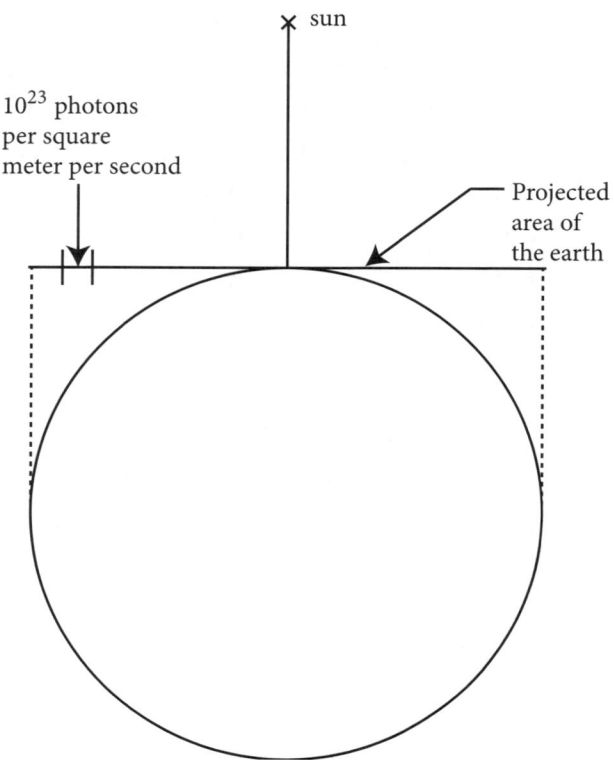

Most of the photons are reflected but many are absorbed by the earth and are then re-emitted. Most of the photons we see are those that have been reflected.

A single photon-emitting channel from one communication satellite, along with relay satellites, continually covers the earth with photons. These photons are intercepted by our radios and televisions which then generate sound waves for our ears and other photons for our eyes.

Just what is a photon? Since everything is made of brutinos

we know a photon is made of brutinos. Consider a hydrogen atom at rest with its electron in its lowest chemical state. What this means is that the electron has its smallest orbital radius. Also, the electron's orbital velocity is approximately the speed of light divided by 137, or $2.19 \times 10^6 m/s$. Without specifying what a photon is or where it comes from, for the present, let a photon (coming at the speed of light, of course) impinge on the atom. What will occur is the atom will capture part of the photon and scatter off part as a *lower energy* photon and the hydrogen atom will be accelerated.

The hydrogen atom will be heavier than before impact. In fact, the mass of the hydrogen atom grows as a function of its velocity as given by

$$m_v = \frac{m_o}{\sqrt{1-(v/c)^2}}$$

Where m_o is the atom rest mass, m_v is the mass at velocity v and c is the speed of light. This equation clearly shows that an atom cannot exceed the speed of light – which is an obvious result of trying to get something moving faster than the impactors. We show a plot of m_v/m_o versus speed in speed-of-light units.

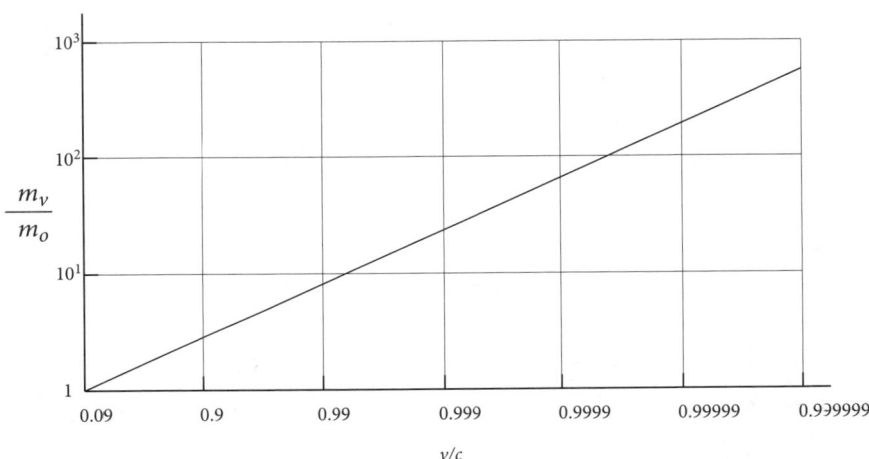

Just what is the nature of the impactor, i.e., the photon? Photons are clouds of brutinos strung out uniformly over one sine wave of length. Usually we show a sine wave as below.

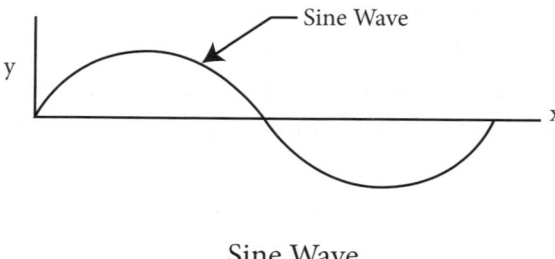

Sine Wave

However a complete sine wave can begin at any *x* location as we show below.

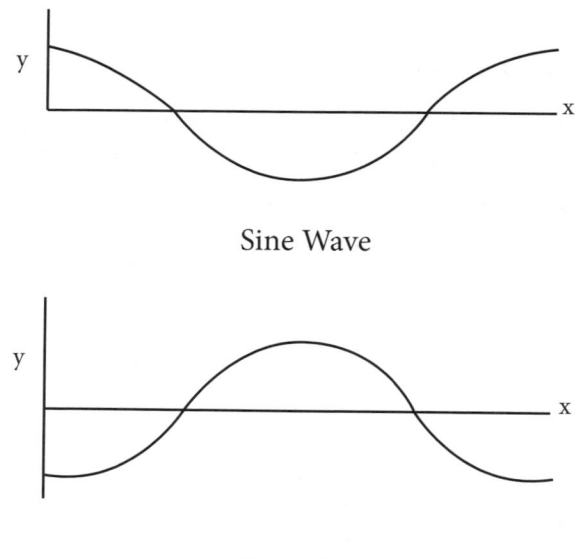

Sine Wave

Sine Wave

Recall that the photon is immersed in the brutino gas background so the brutinos making the photon are identical to those of the background except for their motion. Photons have some similarity to sound waves. However, in a sound wave there is no transport of mass. Sound waves transmit pulses of energy while a photon actually transports mass with its energy.

The figure below illustrates the mass transport of a photon.

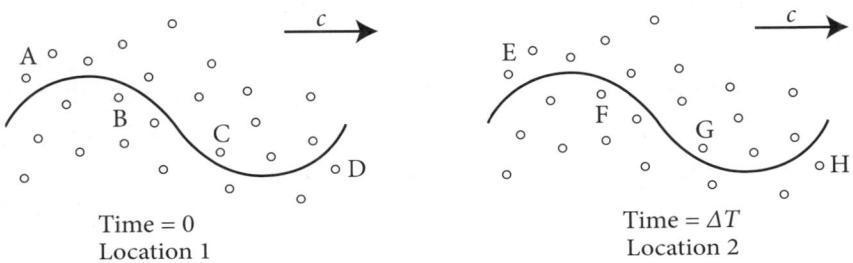

Photon Mass Transport

Particles A, B, C, and D are masses beyond the average density at Locaton 1. At a time later, at Location 1, the extra mass, as noted by particles A, B, C, and D is gone. The extra mass has moved to Location 2 at time Δt later. This extra mass is signified by E, F, G, and H.

We show background particles, unlabeled, which are not flowing. They have an average velocity around ten times the speed of light. Possibly all the transport resides in the number of particles required to make the masses of the photon. We only show four particles but a photon has many, as many as 10^{30} to 10^{40} particles. An instant later there are still almost the same number of particles moving with an average value of the speed of light relative to the *rest* background, but the particles are not the same ones. The transport particles still have the sine wave distribution of location. With the above discussion we note that this is our best estimate of the structure of a photon.

Let us return to the acceleration of matter. Instead of a hydrogen atom let us accelerate a proton, which is a simpler piece of matter. As a result of the proton's neutrino taking a circular path and stirring the background gas, there is a pulsing radial wave and a traveling tangential wave of motion of the background gas. This wave pattern is the proton's electrostatic field. The photon intercepts this field and forms a closed ring of mass around the proton. The circumference of this ring is longer than the wave length of the impinging photon since part of the photon was scattered. The smaller the mass of a photon the longer its wavelength is. Another photon, also of longer wave length, scatters from the proton. As a result of the impulse provided by the scattered photon and the captured mass the proton is accelerated. The figure shows this phenomenon.

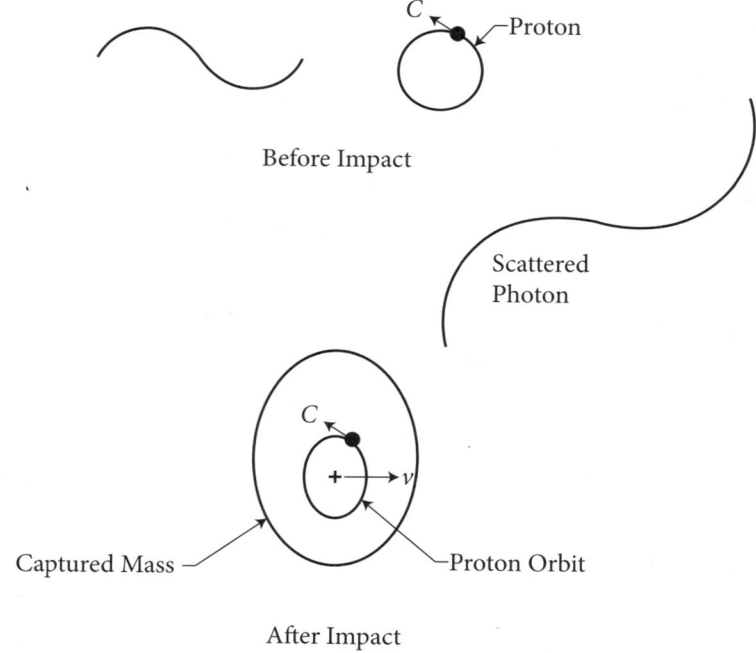

Compton Scattering

This phenomenon is known as *Compton Scattering* named for the discoverer A.H. Compton. Dr. Compton did not specify how this captured mass was located in the proton.

The ring of mass is captured *off center* so that the charge oscillates up and down as the proton moves. This up-down motion is the wave property which matter has when it moves and produces the magnetic field.

When the proton stops, it sheds the mass by unwinding it. This process produces the sine wave distribution of mass which is the photon.

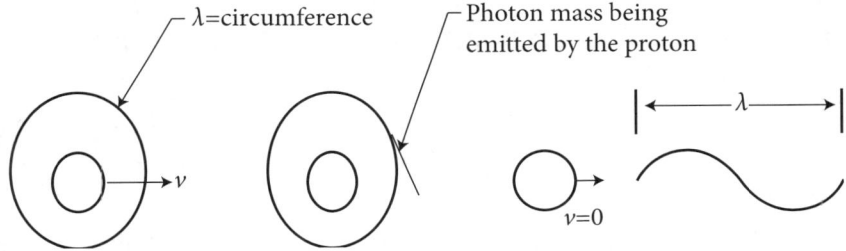

Photon Emission

The photon does not undulate as it moves. It acts like a piece of wire bent into a sine shape and moves much as a rigid body. The energy of a photon is its mass time the square of the speed of light. The energy E also is related to its frequency by

$$E = h\nu$$

where h is Planck's constant per cycle $h = 2\pi\hbar$ (where $\hbar$ is Planck's constant per radian) and ν is the cycles per second. If it takes 10^{-6} seconds for a photon to pass a person at rest then ν is $1/10^{-6}$ or 10^6 cycles per second. Since

$$\nu = \frac{c}{\lambda}$$

The energy is

$$E = h\nu = hc/\lambda$$

where λ is the photon wave length. Thus, we see that the longer the wave length the less the energy and the less the number of brutinos in the photon. This is shown by the equation

$$E = Nmc^2 = hc/\lambda$$

where N is the number of brutinos in the photon and m is the mass of each brutino.

$$N = [h/(cm)]/\lambda$$

We show a schematic view of a high energy-short wave length photon and a low energy-long wave length photon.

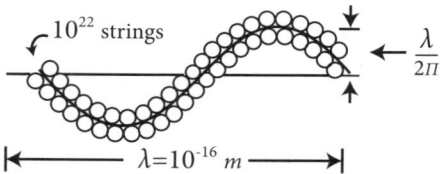

a. Short Fat Photon

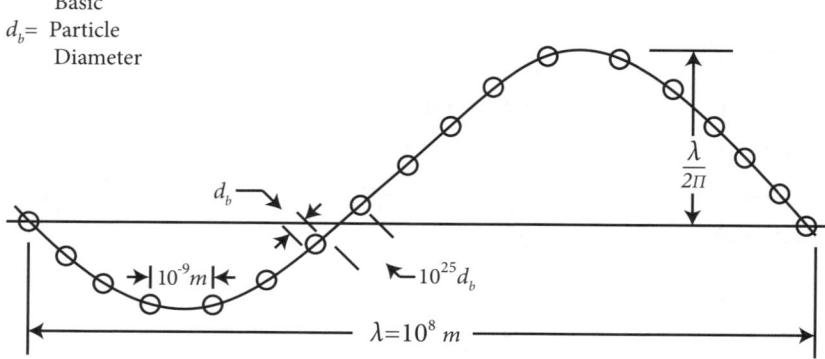

b. Long Skinny Photon

Photons Move Almost Everything We See

Photons lose energy as they travel in space. This loss is a direct result of the conservation of angular momentum. Angular momentum conservation is exemplified by a speed govenor, see the figure.

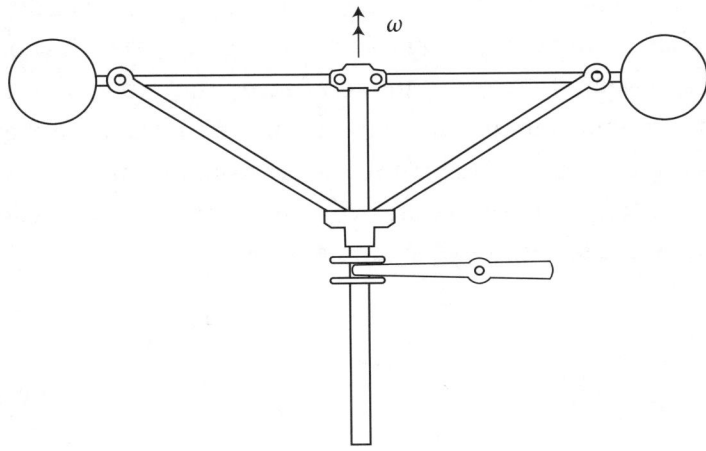

Turning Slowly

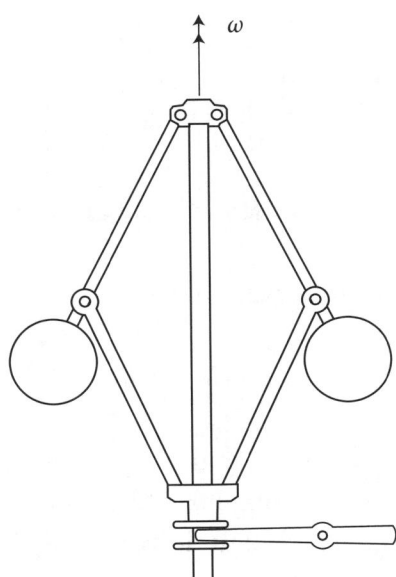

Turning Faster

Angular momentum is *mass* × *distance* from *center of rotation* × *velocity*. The angular momentum for the left figure is that due to the body plus the arms. With the arms retracted, the distance from the mass center to the center of rotation reduces so that the angular velocity increases. The mass in the atom which is to be emitted as a photon has an angular momentum whose value is Planck's constant, h ($6 \times 10^{-34} kg\text{-}m^2/s$). As the photon leaves the atom, the angular momentum taken about the atom center continues to be h even as the photon mass continually moves away from the atom.

The left figure shows the hydrogen atom before emission and the right figure shows the atom and photon after emission

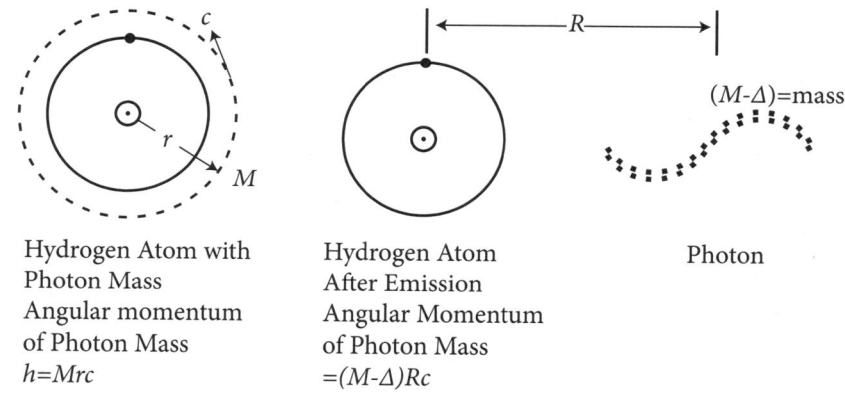

Hydrogen Atom with Photon Mass
Angular momentum of Photon Mass
$h = Mrc$

Hydrogen Atom After Emission
Angular Momentum of Photon Mass
$= (M-\Delta)Rc$

Photon

Angular Momentum of Photon Mass Before and After Emission

Angular momentum conservation gives

$$Mrc = (M - \Delta)Rc$$

Now as R increases Δ increases and, thus, the photon mass decreases. The decrease in mass is accomplished by removing one brutino for each cycle – which is the smallest decrease possible.

Based on this analysis and the assumption of one brutino

removed for each wave length of travel from the most distant star to the earth gives the brutino mass as $10^{-66} kg$.

The foregoing analysis indicates why we can only observe stars out to a distance of 10^{10} light years, or 10^{26} meters. The description of this phenomenon, and the description of the origin of matter of this theory are in sharp contrast with the currently prevailing theory that the universe began as an infinitely large mass density in an infinitely small volume and began expanding 10^{10} years ago. With the latter concept, stars that are closer are moving away from the earth slower than more distant stars. The amount of the shift in frequency from the frequency of a photon emitted on the earth is the result of the motion of the star. The frequency shift due to motion is familiar for a train whistle as a train is moving away from an observer. However, according to our theory the photon just loses its mass as it travels. Distant stars are not moving away from the earth and they emit the same energy photons as close stars.

When matter at rest is impacted by a photon the neutrino making up the matter takes a two-dimensional spiral path as shown in the following figure.

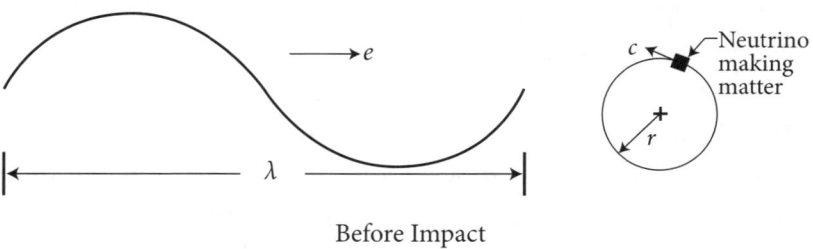

Before Impact

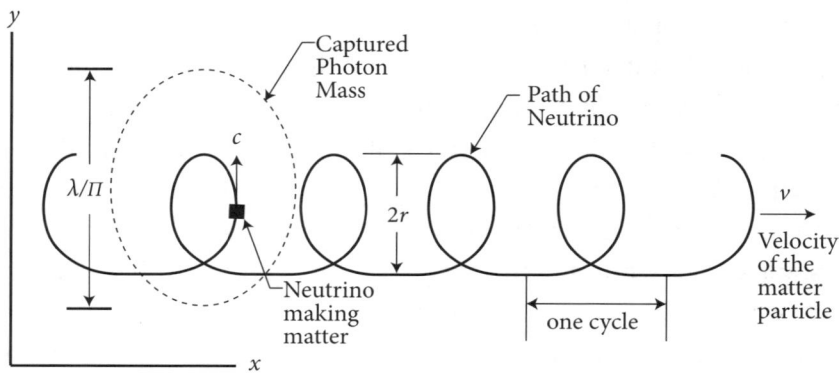

After Impact

After impact it takes longer for the neutrino to travel one cycle. The circular travel of the neutrino at velocity c, as a result of the impact, becomes a plane spiral path and gets the matter particle moving at velocity v. The time for a cycle at velocity is T_v and is related to the rest time T_o by

$$T_v = \frac{T_o}{\sqrt{1-(v/c)^2}}$$

If the observer of the neutrino motion were to get on a vehicle moving at the same speed and in the same direction as the matter particle, the neutrino path would become an ellipse.

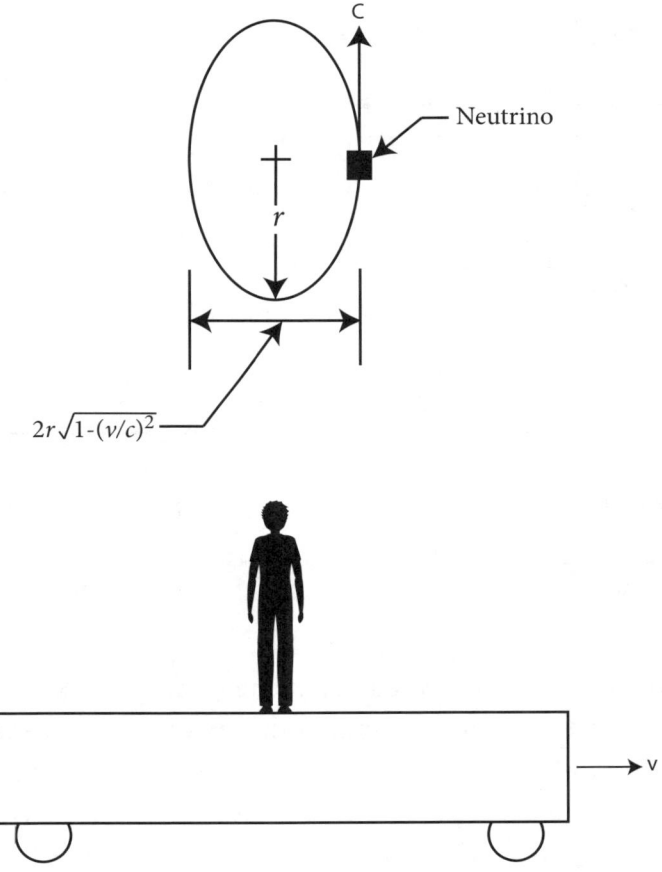

The observer on the cart moving at velocity v sees the neutrino moving in an ellipse with a major diameter of $2r$ and a minor diameter of $2r\sqrt{1-(v/c)^2}$. Thus, when matter moves, its dimension decreases by the factor $\sqrt{1-(v/c)^2}$. – since every matter particle experiences this length reduction.

As another example of an interesting effect of time clocks running slower and matter shortening when the clocks and matter are moving is that the velocity of light measures the same independent of the frame from which the measurements are made. Consider the two *imaginary* tape measures, one of which is 2×10^8 meters long and the other which is 4×10^8 meters long.

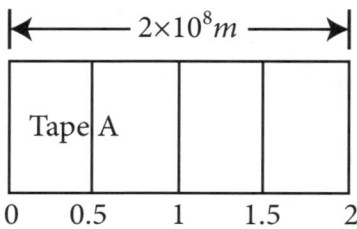

Now, let tape A remain at rest and let tape B move at 0.866 times the velocity of light c. Note now that both tapes have the same length since $\ell_v = \ell_o\sqrt{1-(v/c)^2} = \ell_o\sqrt{1-(.866)^2} = \ell_o(0.5)$. $\ell_v = 4(0.5)\times10^8 = 2\times10^8 m$

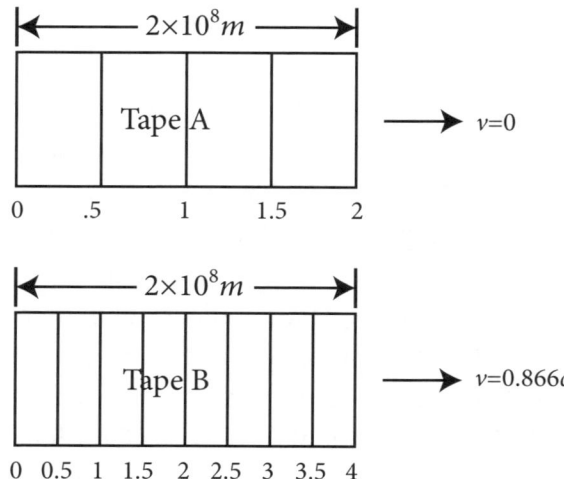

Photons Move Almost Everything We See

Let us now wrap these two tapes around flat circular disks – one of which is at rest and the other one rotating so that the peripheral velocity is 0.866*c*.

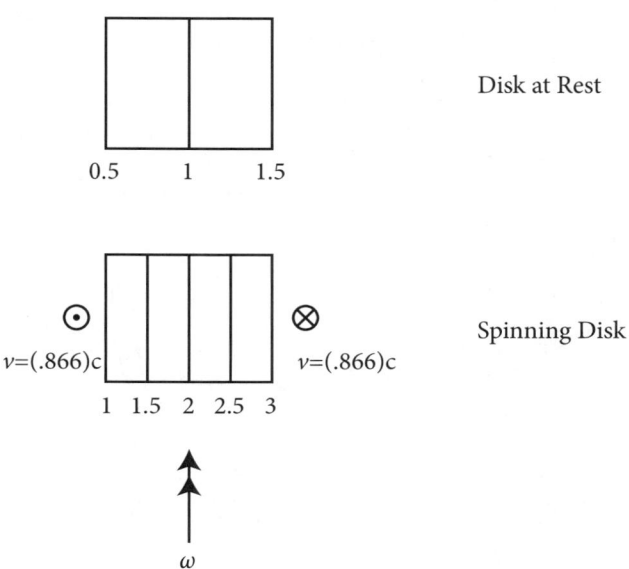

Let us triple the size in the following figure of the circular disks and surround them with mirrors which will cause a photon to go around the disks as shown in the next figure. The following figure shows a view looking down on the disks.

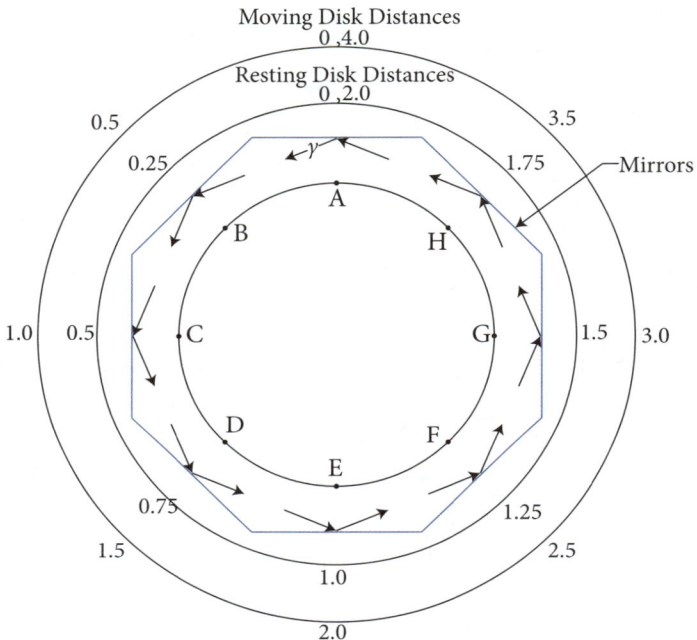

A light source is at A, and it emits a photon which, because of the mirrors, travels the path ABCDEFGHA.

Next, let's put clocks on both disks and record the time to travel ABCDEFGHA.

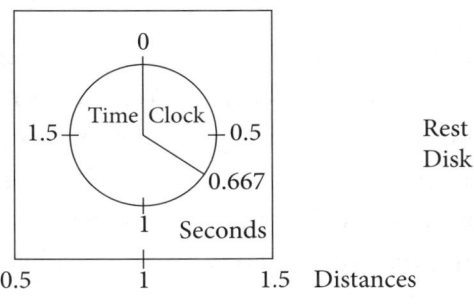

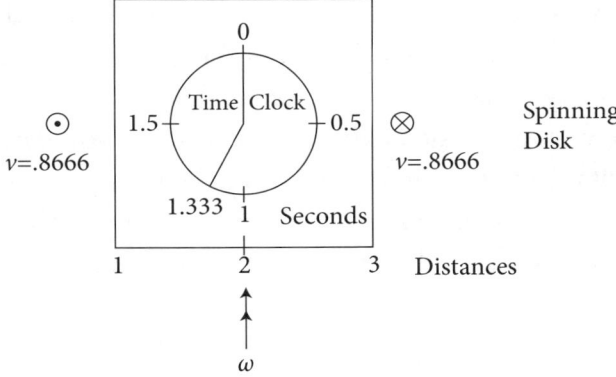

We note that the clock on the disk which was not rotating recorded 0.667 seconds and the clock on the rotating disk recorded 1.333 seconds.

The velocity of the photon as measured by the distance on the resting disk and the elapsed time on the clock is

$$v_{light(rest\ frame)} = \frac{\text{Measured Distance}}{\text{Elapsed Time}} = \frac{2 \times 10^8}{0.667} = 3.00 \times 10^8 m/s = c$$

The velocity of the photon as measured by the rotating disk and moving clock is

$$v_{light(moving\ frame)} = \frac{\text{Measured Distance}}{\text{Elapsed Time}} = \frac{4\times10^8}{1.333} = 3.00\times10^8 m/s = c$$

We note that the measured photon velocity is the same for both cases. In the real world it is not possible to have such long tapes and such a fast rotating disk, but actual experiments verify the above result. Further, up until now there has not been a method for discerning a rest frame. Thus all inertial[1] frames are equivalent since they all give the result that the velocity of light is $3\times10^8 m/s$ relative to a frame no matter what its velocity is.

Recently a method was devised to determine the rest frame – i.e., a frame which is not moving with respect to the ether. This method was based on the mass growth of matter as a function of its absolute velocity. If matter were accelerated in a direction opposite the absolute velocity of the earth then its mass growth would be less than a simultaneous experiment run in the exact opposite direction. A tenuous determination was made that the Earth's absolute velocity is somewhere around a half percent of the speed of light.

---

1 An inertial frame is one which is neither rotating nor being accelerated in any direction.

11. Summary

Brutinos are the smallest things in the world. Everything is made of brutinos and nothing else. The brutinos are gas particles which make up the gaseous ether which pervades the universe. Because of the random motion of the gas particles, winds are formed in the gas. Infrequently the winds can form a tornado-like flow pattern. Some of these tornadic flows produce condensations of the background that are stable and translate slowly (at 8% the average particle speed). These condensations are neutrinos and they populate the universe. The neutrinos all travel at the speed of light, have the same angular momentum, and are formed with a broad range of mass.

A neutrino with the mass of a proton can get knocked into a circular path with its thrust directed through the center of the path. That configuration is a proton. Simultaneously with the formation of the proton, the electron is formed to counterbalance the flows produced by the protons. Counterbalancing the flows necessitates that the electron be formed with the mass it has and the electron orbital velocity is specified by the flows produced by the proton. Basically, the flows during the formation of the proton act as a template for the formation of the electron.

The neutrino making the proton goes around in a circle and stirs the background gas. This stirring of the background is the electrostatic field. The electron similarly stirs the background to make the negative electrostatic field. The fields interact resulting in an attractive force binding the electron to the proton to make the hydrogen atom.

The negative electrostatic field of the electron mixed with the positive electrostatic field of the proton in the hydrogen atom produces a residual field which is the gravitational field. Every hydrogen atom has a gravitational field and two hydrogen atoms will attract each other. Since hydrogen atoms are plentiful throughout the universe and since they are made continually, the hydrogen atoms begin to assemble into clusters. Over the ages theses clusters grow and eventually form hydrogen stars.

Large hydrogen stars have strong gravitational fields. These fields can attract a hydrogen atom strongly enough so that the atom will have enough energy when it reaches the star to decrease the electron radius by a factor of 100,000 and transmute hydrogen into a neutron.

Neutrons and protons are nucleons. The proton-sized neutrinos in nucleons when placed side-by-side can attract each other and fuse to make nuclei of larger atoms. As stars grow they make more and more large atoms.

As a star continues to grow the internal gravitational forces get large enough to collapse the electron orbits of the atoms and transmute them into neutrons – to produce a neutron star.

As a neutron star continues to grow its gravitational field becomes so strong that the nuclear orbital structure collapses. The star then explodes, producing all sizes of matter. Some chunks are large and produce hydrogen stars, some are small and eventually cool down to make planets, moons, asteroids, and other space junk.

All photons lose energy as they travel – they lose one brutino for each wave length of travel. Photons emitted from hydrogen, as from a hydrogen star, can travel 10^{26} *meters* before losing all their energy. Thus, we on the earth can only see events inside a sphere with a radius of 10^{26} *meters*.

Bibliography

Brown, Joseph M., *The Grand Unified Theory of Physics*. ISBN: 978-0-9712944-6-2, Basic Research Press, Starkville, MS, 2004.

Brown, Joseph M., *Photons and the Elementary Particles*. ISBN: 978-0-9712944-5-5, Basic Research Press, Starkville, MS, 2011.

Brown, Joseph M., *The Neutrino*. ISBN: 978-0-9712944-7-9, Basic Reasearch Press, Starkville, MS, 2012.

Brown, Joseph M., *Foundations of Physics*. ISBN: 978-0-9883180-0-7, Basic Research Press, Starkville, MS, 2012.

The majority of the results in this book have been derived and presented in *Foundations of Physics*.

Index

A
action at a distance 79
angular momentum 68, 70, 72, 74, 75, 121, 122, 131
angular momentum loop 74
asteroids 110, 132
atoms vi, 6, 7, 87, 90, 91, 92, 93, 96, 97, 98, 100, 101, 102, 104, 132

B
background 20, 21, 37, 42, 44, 45, 70, 71, 72, 73, 82, 87, 97, 116, 117, 118, 131
big bang 63, 133
Big Bang 107
Big-Big Bang 101, 107
Bohr 75
Brown i, ii, iii, 133
brutino i, ii, iii, vi, 22, 26, 35, 36, 37, 46, 54, 84, 90, 116, 120, 122, 123, 132
brutinos 24, 26, 28, 31, 35, 36, 37, 42, 44, 46, 56, 73, 87, 92, 114, 115, 116, 120, 131

C
circulation 62, 70
collision 32, 33, 34, 35, 36, 46
Compton Scattering 118
condensation 56, 58
copper 6, 7
core 44, 46, 54, 55, 56, 57, 58, 59, 60, 61, 62, 70
coupling velocity 86
cylindrical 43, 46, 62

D
defining sphere 54, 61
density 26, 36, 37, 63, 117, 123
detect 113

E
earth 1, 29, 30, 112, 113, 114, 123, 132
earth's absolute velocity 130
electron vi, 1, 8, 9, 10, 11, 14, 72, 73, 74, 75, 76, 78, 79, 81, 82, 83, 84, 85, 86, 88, 89, 90, 95, 98, 99, 115, 131, 132
electron formation 82
electrostatic 73, 77, 79, 81, 84, 85, 86, 87, 90, 95, 118, 131
ellipse 124, 125
energy 35, 41, 44, 59, 96, 115, 117, 119, 120, 121, 132
explosion 111

F
fine structure constant 86
free molecular 61
free-molecular-flow 58
fuse 132

G
gas vi, 22, 25, 26, 28, 36, 37, 44, 46, 48, 58, 63, 70, 72, 77, 80, 82, 83, 87, 113, 116, 118, 131
glue iv, 77
God particle 28
govenor 121
gravitation 90, 92, 100
gravity vi, 87, 90, 91

Index

H
head-on 32
Higgs 28
hydrogen vi, 7, 8, 75, 81, 82, 84, 87, 90, 91, 92, 93, 94, 95, 96, 100, 101, 102, 104, 115, 117, 122, 131, 132
hydrogen star 93, 132

I
inertial frame 130
inertial loop 76
in-phase 79, 80, 84, 85

L
left-hand 78
Ludwig Boltzmann 41

M
magnification 5, 6, 20, 21
magnify 2, 4
magnifying glass 2
mass 1, 23, 26, 35, 36, 37, 43, 63, 64, 68, 70, 75, 76, 81, 87, 90, 96, 100, 115, 117, 118, 119, 120, 122, 123, 131
mass growth 130
Maxwell 41
Maxwell-Boltzmann 41
mean free path 26, 37, 57, 58
Mean free path 55, 56, 57
mean speed 28, 41, 44, 46
most probable 40, 41
move 37, 39, 42, 44, 86, 126

N
neutrino vi, 1, 12, 19, 20, 37, 41, 42, 44, 45, 46, 48, 49, 50, 51, 52, 54, 55, 56, 57, 59, 61, 62, 64, 67, 68, 70, 71, 73, 76, 77, 78, 79, 81, 82, 83, 84, 85, 86, 87, 88, 95, 117, 123, 124, 125, 131
neutron vi, 90, 95, 96, 100, 104, 107, 110, 132
neutron star vi, 104, 107, 110, 132
neutron star explodes vi
normalized 40
nuclear bomb 97
nuclear force 96
nuclei 132
nucleon 98

O
orbit 8, 64, 68, 70, 90
out-of-phase 79, 80, 84
oxygen 7

P
penny 3, 4, 5, 6
perfectly elastic 31, 36, 37, 63
photon 1, 114, 115, 116, 117, 118, 119, 120, 122, 123
planet 111
planets 110, 112, 132
polarity 73
probability 41, 54, 75, 86
proton vi, 1, 8, 9, 10, 11, 16, 64, 68, 69, 70, 72, 73, 75, 76, 77, 78, 79, 80, 81, 82, 83, 84, 85, 86, 87, 89, 90, 95, 96, 97, 98, 99, 117, 118, 119, 131, 132
proton formation 73

R
rare 27, 41
resonance 84, 85, 86
resonant 84
right-handed 70, 73
rms speed 28, 41, 42

135

root mean square 28
rotational vector 16

S
semi-solid 46, 58
skater 69
solid core 44, 54, 58
space 24, 28, 30, 36, 37, 46, 63, 86, 96, 110, 121, 132
spacing 23, 25, 29, 36
star vi, 93, 96, 104, 107, 110, 111, 123, 132
strong nuclear force 96
subsonic 61
supersonic 61

T
template 72, 73, 82, 131
three-dimensional 30, 34, 63
thrust 14, 42, 59, 67, 68, 131
tornado 46, 48, 131
transmute vi, 132
traveling wave 77, 78, 88
twist 16, 73, 78, 79, 81
two-dimensional spiral 37, 123

U
universe vi, 22, 23, 24, 27, 29, 30, 41, 46, 59, 63, 92, 123, 131, 132

V
velocity 16, 32, 33, 34, 35, 36, 37, 38, 39, 40, 41, 42, 44, 45, 67, 70, 83, 84, 86, 115, 117, 122, 124, 125, 131

W
water 7, 79
winds 48, 63, 131
world vi, 1, 131